Risky Cities

Nature, Society, and Culture

Scott Frickel, Series Editor

A sophisticated and wide-ranging sociological literature analyzing nature-society-culture interactions has blossomed in recent decades. This book series provides a platform for showcasing the best of that scholarship: carefully crafted empirical studies of socio-environmental change and the effects such change has on ecosystems, social institutions, historical processes, and cultural practices.

The series aims for topical and theoretical breadth. Anchored in sociological analyses of the environment, Nature, Society, and Culture is home to studies employing a range of disciplinary and interdisciplinary perspectives and investigating the pressing socio-environmental questions of our time—from environmental inequality and risk to the science and politics of climate change and serial disaster, the environmental causes and consequences of urbanization and war making, and beyond.

For a list of all the titles in the series, please see the last page of the book.

Risky Cities

The Physical and Fiscal Nature of Disaster Capitalism

ALBERT S. FU

Rutgers University Press
New Brunswick, Camden, and Newark, New Jersey, and London

Library of Congress Cataloging-in-Publication Data
Names: Fu, Albert S., author.
Title: Risky cities: the physical and fiscal nature of disaster capitalism / Albert S. Fu.
Description: New Brunswick, NJ: Rutgers University Press, [2022] | Series: Nature, society, and culture | Includes bibliographical references and index.
Identifiers: LCCN 2021031064 | ISBN 9781978820302 (paperback) | ISBN 9781978820319 (cloth) | ISBN 9781978820326 (epub) | ISBN 9781978820333 (mobi) | ISBN 9781978820340 (pdf)
Subjects: LCSH: Cities and towns. | Urbanization. | Environmental disasters. | Risk management.
Classification: LCC HT151 .F79 2022 | DDC 307.76—dc23
LC record available at https://lccn.loc.gov/2021031064

A British Cataloging-in-Publication record for this book is available from the British Library.

Copyright © 2022 by Albert S. Fu
All rights reserved

No part of this book may be reproduced or utilized in any form or by any means, electronic or mechanical, or by any information storage and retrieval system, without written permission from the publisher. Please contact Rutgers University Press, 106 Somerset Street, New Brunswick, NJ 08901. The only exception to this prohibition is "fair use" as defined by U.S. copyright law. References to internet websites (URLs) were accurate at the time of writing. Neither the author nor Rutgers University Press is responsible for URLs that may have expired or changed since the manuscript was prepared.

♾ The paper used in this publication meets the requirements of the American National Standard for Information Sciences—Permanence of Paper for Printed Library Materials, ANSI Z39.48-1992.

www.rutgersuniversitypress.org

Manufactured in the United States of America

For my daughter, Lara

Contents

	Abbreviations	ix
	Introduction	1
1	Living with Disaster and Capitalism	16
2	Sinkholes and the Risky Foundations of Cities	38
3	The Logistical Nightmare of Trash and Urban Nature	57
4	Fire, the Wildland–Urban Interface, and Feedback Loops	77
5	Assessing and Managing Risk	95
	Conclusion: Regenerative Urbanism	115
	Acknowledgments	125
	Notes	127
	Index	165

Abbreviations

ACS	Actividades de Construcción y Servicios, S.A.
ADB	Asian Development Bank
BOO	build-own-operate model
BOT	build-operate-transfer model
CAPRA	Central American Probabilistic Risk Assessment
CAT	catastrophe bond
CCRIF	Caribbean Catastrophe Risk Insurance Facility
CME	Chicago Mercantile Exchange
EIB	European Investment Bank
EIOPA	European Insurance and Occupational Pensions Authority
EPA	U.S. Environmental Protection Agency
ESG	environmental, social, and governance
FAO	Food and Agriculture Organization of the United Nations
FEMA	Federal Emergency Management Agency
GFDRR	Global Facility for Disaster Reduction and Recovery
GNDR	Global Network of Civil Society Organizations for Disaster Reduction
ILS	insurance-linked securities
IMF	International Monetary Fund
IPCC	Intergovernmental Panel on Climate Change
ISO	International Standards Organization
LEED	Leadership in Energy and Environmental Design

Introduction

• •

> Man made himself *master* and *possessor of nature*, of the sensible, of substance. It was throughout this that he divided himself against himself, in realizing himself.
> —Henri Lefebvre, *Rythmanalysis: Space, Time, and Everyday Life*

In fall 2019, Pacific Gas & Electric (PG&E), one of the largest utility companies in California, announced that it would be shutting off power to millions of residents. As warm, dry winds picked up, the goal was to avoid starting a wildfire as PG&E did a year earlier in Butte County in Northern California. The 2018 Camp Fire, as it is now called, was caused by a transmission line igniting dry brush, and it would become the costliest disaster in the world that year. The fire infamously engulfed the town of Paradise, burned over 150,000 acres, and led to at least ninety fatalities. The whole region was affected by smoke and ash, and the fire also led to the most extensive hazardous material cleanup in California's history. Communities dealt with record levels of disaster debris as well as the contamination of local water supplies. Charred earth also contributed to mudslides, flash floods, and sinkholes in the months following the firestorm. Far from being a stand-alone crisis, the fire created a series of new disasters.

The story of PG&E, the Camp Fire, and its aftermath is interesting not only because of its scale but also because it is an example of how disaster risk is a systemic phenomenon. In addition to the chain of physical disasters caused by the fire, we must examine the social, economic, and political context in which they occurred. PG&E's history is grounded in providing gas and electric power to San Francisco, which allowed both company and city to expand throughout the twentieth century. Today, the company provides energy from Humboldt County in the northern part of the state all the way down to Santa Barbara, just north of Los Angeles. For this to occur, the company created and operates an extensive system of potentially dangerous powerlines to keep California's cities functioning and growing. While anthropogenic climate change is a significant influence on wildfires in the region, urbanization and the accompanying transformation of landscape also cause wildfires. Thus, we see that the expansion of human settlements is a significant driver of disaster risk.

On a warming planet with increasingly chaotic weather, cities are increasingly depicted as risky places. According to the World Bank, the severity of both natural and human-made disasters is on the rise, resulting in death and human suffering. Hurricanes, storms, wildfires, extreme heat, subsidence (sinkholes), and the ever-looming threat of massive earthquakes pose increasing threats to human settlements. Average mortality and death per catastrophe increased between 2006 and 2015.[1] British reinsurance firm Aon Benfield estimates that the annual financial cost of natural disasters hitting cities will increase from $250 billion to $314 billion by 2030.[2] As a result of climate change, there is no doubt that cities occupy an increasingly risky position.

The increasing prevalence of risk on a warming planet is also a result of the interconnected nature of our cities. Cities are simultaneously significant nodes within local infrastructure systems as well as the world economy. Take PG&E and California's network of powerlines, which link human settlements throughout the region. The 2018 Camp Fire illustrates the interconnected nature of risk and the social domino effect of disaster. A single catastrophe has the potential to set off a chain reaction within and between cities. Fire can create logistical problems for waste and lead to mudslides, flash floods, and sinkholes. Risks also include breakdowns in transportation, utilities, and local and global financial networks.[3]

With over half of the world's population living in urban areas, it is ever more critical that we think about how cities exist in a world that is warming and in which the costs associated with disaster are increasing. Driving increased disaster risk is urban expansion or growth, which places people and buildings in precarious environments. Wildfire is a prime example: People are moving into dry and flammable areas. This obsession with expansion often leads to inadequate—if not dangerous—management of resources and waste flows, causing human health emergencies, drought, and famine around the world. All these problems serve as tinder for disaster. At the same time, construction, pollution, and the exploitation of resources disrupt ecosystems, creating new threats.

Yet we keep building and believing we can engineer ourselves out of this risky situation. As this introduction's opening quote by French sociologist Henri Lefebvre suggests, humans have continuously attempted to conquer nature to meet the demands of ongoing capitalist expansion and growth. As Lefebvre puts it, "There is nothing which is not acquired and won from nature."[4] Knowledge, technology, and urbanization are tools humans have used to change the physical form of the environment around them. In turn, we can understand urbanization as an expansion-oriented technology to conquer nature. It is a tool that transforms the natural landscape in the interests of capitalism to accumulate profit. Capitalism is not just about making a profit; instead, it involves increasing profit margins by continuously expanding markets. This logic of expansion drives urban growth. Humans have overcome past limitations to city-building by moving earth, reclaiming land from the sea, running infrastructure to inhospitable regions, and developing new materials that allow cities to expand outward and upward. None of this is new. However, under contemporary capitalism, the deployment of new technological and financial strategies radically breaks down the historic (and natural) relationships among people, nature, and urban spaces.

Urbanization does not occur without blowback and negative repercussions. Cities are increasingly pressed to develop new strategies to deal with environmental *bads*—such as disaster—both fiscally and physically. For instance, we hear buzzwords such as adaptation, mitigation, and resilience in construction, building management, and urban planning. New technologies are budgeted for, invested in, and deployed regularly. Techniques include innovation in finance that distribute the costs associated with catastrophe throughout financial markets and society. At the same time, despite an increase in insured losses from disasters in recent years, the insurance industry remains highly profitable. For example, Aon Benfield has noted that global reinsurer capital has steadily grown from $340 billion in 2008 to $600 billion in 2017.[5] Here, we see a contradiction in which profit is dependent not just on the catastrophe itself but also the threat of disaster.

The redistribution of risk means that short-term fixes to disaster risk are more prevalent than long-term changes to city-building. This leads to more disaster or blowback. These are the unintended consequences of policies and strategies that fail to address the social, political, and economic drivers of so-called natural hazards. Again, this is tied to growth-oriented urban policy and the financialization of disaster mitigation, whereby "solutions" that receive the most support are also those that generate the most profit for shareholders. An example of this ideology is how American policymakers and economists use new home sales as a measure of economic growth. On a global level, we use the level of urbanization as a measure of development. These two indicators are examples of how the logic of the urban growth machine permeates our thinking.[6]

Risky Cities is about how urban regions have commodified environmental hazards, which in turn creates new problems and risks for society. This book is a critical examination of how environmental *bads* such as disaster become valuable *goods* through disaster capitalism and urban development. *Risky Cities* is a look at how contemporary capitalism challenges urban sustainability and well-being. It is increasingly evident that elites profit from the building and rebuilding of cities. The need for fundraising is common after a disaster. This happens through the allocation of government funds, external aid, and the issuance of credit for rebuilding. Politically, the raising of funds often includes the diversion of resources to support disaster relief. Unfortunately, much of the money raised ends up supporting private companies. Contractors are typically hired to handle post-disaster logistics. As such, it is not surprising that the critiques of post-disaster responses are many. Naomi Klein has popularized the notion of disaster capitalism, referring to how elites in power benefit from ongoing crises. Klein and other scholars have since looked at urban crisis and profiteering. This work vividly describes how our capitalist system thrives on disaster through the rise of security firms, post-disaster opportunism, and the creation of policies that "shock" the economy.[7] For example, John Mutter has compared post-disaster aftermaths to economic development programs. Mutter and others argue that such programs, imposed by external actors, typically result in exploitative policies that undermine local sustainability to generate profit while creating new threats.[8]

However, *Risky Cities* is not merely about post-catastrophe profiteering or a critique of myopic policymaking. Instead, I adopt a historical and global approach to how cities live and profit from the legacy and ongoing threat of sinkholes, garbage, and fire. Cities are not merely dots on a map; instead, they operate within an extensive global system that connects and distributes risk. The built environment, therefore, is a product and consequence of disaster or the threat of catastrophe. *Risky Cities* is an examination of how environmental *bads* are deeply embedded in the structure of modern cities and capitalism. Given the interconnected and interdependent nature of modern cities, it is necessary to consider disasters from both historical and global perspectives.

What Is a Risky City?

Are cities riskier and more likely to experience a disaster than nonurban areas? Cities have always coexisted with the risk of catastrophe. There is no doubt that high population density means greater exposure when disaster strikes. A volcano famously destroyed Pompeii in 79 C.E. The Lisbon earthquake of 1755 and the San Francisco earthquake of 1906 are other well-known catastrophes. We can also include human-made disasters, such as the Great Fire and Great Stink of London in 1666 and 1858, respectively; the Great Chicago Fire of 1871; and the burning of the Cuyahoga River in Ohio in 1969. Given this history, we can see that disasters are a fundamental part of urban life and history.

In our current urbanized and capitalistic world, the situation is far from improved. Without a doubt, urbanization and the expansion of human settlements into new areas has increased exposure and vulnerability to disaster. For instance, energy is produced in one place and consumed elsewhere. As the story of PG&E shows, the threat of wildfire is linked to vulnerabilities in the built environment across an entire region. By increasing interconnectedness and mutual dependence, breakdowns in infrastructure in one city threaten many other cities, both socially and economically. I argue that risk is now deeply embedded within the built environment thanks to contemporary capitalism's financialization of every sphere of human life. That is, everything is reduced to something that can be traded in a market—even disaster. Moreover, while everything is subject to risk, specific weak points are a by-product of fiscal decisions by actors who privilege expansion over addressing real dangers and threats.

Ulrich Beck famously coined the term "risk society" in the 1980s to discuss how modern society organizes itself in response to "hazards and insecurities induced and introduced by modernization itself."[9] According to Beck, the contemporary world is increasingly exposed to hazardous by-products of modernization, despite our reliance on science. Today, fear and uncertainty have a significant influence on decisionmaking and the structure of society. Yet technocratic policymakers favor capitalistic expansion over structural solutions to real threats. This is not just about making a profit; it is also about increasing the speed and scale at which revenue is generated. In turn, cities find themselves in a contradictory position of needing to protect themselves from disaster while encouraging expansion. This means that attempts to adapt, mitigate, or build resilient cities are problematically intertwined with capitalism's tendency to expand aggressively.

Risk and disaster are embedded in our society—particularly in places like cities. However, this raises a question: What is "risk"—or specifically, "disaster risk"? Ilan Kelman, examining the historical use of the term in organizations such as the United Nations, has found that "disaster risk" generally refers to "(1) possible losses from a hazard; or (2) potential adverse consequences in a disaster."[10] While such agencies make a qualitative assessment that risk is bad, in the financial world, risk is understood differently. Take, for instance, the rise of the risk-management industry. The International Standards Organization (ISO) is perhaps best known for creating standards that allow for interoperability, interaction, and exchange in a global economy. In recent years, the ISO has defined the term "risk" without such judgments, merely defining it through value-neutral concepts such as "uncertainty" and "unintended." Risk, therefore, can be "positive and/or negative."[11] In other words, risk is not necessarily *bad*; in fact, it is a potential *good*. This understanding comes from risk management as a sector, as well as from the emergence of a disaster-industrial complex, whereby disaster-related industries are integrated throughout our economy. For example, the ISO 31000 guidelines established in 2009 provide a framework for managing risk for various organizations, including businesses and cities. In particular, the ISO advertises its ability to help organizations enhance "the likelihood of achieving their objectives and increasing the protection of their assets."[12] In other words, beyond merely ensuring viability post-disaster, there is also a financial focus.

Despite appearing to be scientifically informed, existing power relations influence international standards. Amy Quark has argued that those who establish rules are not neutral and are highly political.[13] Yet the language and practice of risk management are now hegemonic and institutionalized in our era of global capitalism. For instance, the development of standards such as ISO 31000 as well as disaster-resilience standards such as RELi by the U.S. Green Building Council (USGBC) or the Building Research Establishment Environmental Assessment Method (BREEAM) in the United Kingdom play a role in how institutions manage risk. This often works in tandem with the insurance sector. While private fire insurance existed in the past and influenced the establishment of fire code, companies were not part of global financial services conglomerates intertwined with real estate development and municipal services.[14] As such, in our current political and economic context, cities grapple with risk by integrating it into various financial calculations and fiscal policy.

Addressing the threat of natural disasters on a warming planet is urgent. However, the transformation of risk into a tradable commodity by the financial sector is part of the problem. Financialization turns people, nature, and cities into abstractions, which undermines our ability to build truly resilient cities. Take, for instance, the notion of a city. When thinking of a city, most people think of either a dot on a map or an area contained by a border—which is a conceptualization disconnected from notions of people and nature. In reality, cities are not so easily defined. For instance, Martin Murray has questioned the understanding of a singular city as "self-enclosed, distinctive, discrete, and territorially bound."[15] While Murray is speaking of urbanization as a fragmented and globally interconnected phenomenon, his remarks could equally refer to disasters. Similarly, disaster is not easily defined, its impact is difficult to calculate and its consequences are not limited to well-defined local places.

Cities are made up of different communities connected by local infrastructure and are part of global networks. When disaster strikes, the resulting disruption and impact have various local and global consequences. Response teams and rebuilding operate locally in neighborhoods. Yet flooding, for instance, might be regional. This may complicate disaster relief efforts. Flooding can also be a product of a hurricane that has already damaged several cities and even countries. As such, cities and their inhabitants are not subjected to clearly defined singular threats. While cities deal with localized events, they must also increasingly cope with large-scale natural or human-made catastrophes that extend beyond their borders.

Another example of abstraction is how mass media simplifies disaster. Cities have always lived with the threat of catastrophic events and their consequences. It is typically recent catastrophic "events" that have gained attention from the public, policymakers, and the news media. However, the destruction caused by earthquakes, fire, floods, storms, and breakdowns of infrastructure is a by-product of everyday hazards that are less visible. Social marginalization, as well as poor land use, urban planning, and environmental regulation, in addition to anthropogenic climate change, are all disasters that precede a catastrophic "event" but receive far less attention. We ignore such conditions because of the limited scope of historical memory amid the ever-increasing speed of consumer culture and an unwillingness to confront the role of capitalism in creating disaster.[16]

The Physical Nature

When we see disasters from a distance, we are often most shocked by the physical destruction of human settlements. Our focus is directed by the mainstream news and the social media feeds that present images of towering flames, charred buildings, and rubble. No doubt, the physical nature of disaster draws our attention. While all of this is relevant to our understanding of disaster, far less media attention is given to human shortsightedness in urban planning, land use, and waste management that creates risk. It is not just "nature" that creates disaster risk. The act of building a city—extracting resources, sprawl, and waste production—creates environmental hazards.

In other words, we need to consider the physical (as well as economic) foundations of cities. Urban structures sit atop natural spaces that burn, shake, sink, or fill with water. Waste flows under and above cities, undermining public health. Indeed, people have called for building more resilient cities that are prepared to bounce back from environmental, social, and economic threats. There have been warnings that we must adapt to the new reality of a hotter and dryer climate as well as mitigate the harm it may cause. While these approaches to rethinking infrastructure and building have merits, policymakers typically resort to measures that prioritize urban expansion instead of long-term sustainability. The result is a growing environmental footprint and, in turn, a more significant disaster footprint when catastrophe strikes. The rapid development of housing and commercial centers results in shortsighted planning for infrastructure, and little is done to reduce consumption and expansion—the prime drivers of climate change. Consequently, expansion-oriented growth changes the structure of cities and exacerbates existing dangers while creating new ones.

The material conditions run deeper than structures above ground. The idea that "where" and "how" we build cities affect disaster risk appears straightforward. At the same time, physical materials move or must be moved during and after a disaster. Cities must cope with an influx of material waste and debris while existing infrastructure may be damaged or strained. In catastrophes such as the Camp Fire and other mega-disasters, local and state agencies become overburdened by logistical issues when attempting to handle large quantities of debris. As the U.S. Environmental Protection Agency (EPA) notes, "Serious natural disasters may quickly overwhelm community resources. In these circumstances, communities will likely need to hire private disaster debris management contractors."[17] Following hurricanes, soggy carpets, wet drywall, and mold can make restoration dangerous for workers and homeowners. Trash, cars, furniture, lumber, and uprooted trees must be removed. In the case of wildfire, hazardous waste includes contaminated soil, ash, and other charred material. There are only two options for homeowners and communities dealing with debris removal: Either a contractor hired by the government handles it or a private contractor is employed directly (possibly with assistance from insurance providers). Beyond simply collecting debris, the material must be sorted and disposed of. Some landfills can take hazardous material, and some cannot. Because of logjams at landfills, temporary waste sites are often set up, which draws considerable ire from nearby communities.[18]

Rebuilding also creates waste. Demolition and construction debris must be removed, sorted, and disposed of. Ideally, uncontaminated metals and concrete should be recycled. However, that is easier said than done.[19] The same goes for rebuilding following a natural disaster. Construction after a catastrophe leads to a surge in waste from discarded building materials. Again, some of this waste can be recycled—for example, masonry such as bricks and concrete can be crushed into aggregate and reused in road reconstruction or as fill.[20] Unfortunately, this does not always occur. This physical dimension of disaster capitalism illustrates how catastrophe—such as wildfire—is a disaster not just because of the fire itself. Rather, fire is also a waste problem.

Ironically, after a fire, water can create new risks and disasters. For instance, in Paradise, California, pipes were contaminated with benzene—a cancer-causing chemical—following the fire. This is because heat from a fire can melt metal and PVC pipes.[21] Despite California's drought, rain becomes dangerous after a fire. Water that would usually be absorbed by the earth now readily flows down California's hilly landscape thanks to a slick layer of ash. This water collects material as it runs, creating a debris flow as well as mudslides and flash floods. All of this can destroy buildings, roads, and cars, as well as kill people. The deluge can overwhelm natural and human-made water systems, which can then create sinkholes. In 2018, the Woolsey Fire outside of Los Angeles destroyed 1,600 buildings and devasted almost 100,000 acres. In the winter following the fire, rain caused a flow of debris along with a sinkhole. The large sinkhole engulfed an excavator doing work in the area. Similarly, that same year after the Cranston Fire in California, there were mudslides and a sinkhole opened on Highway 243 that was 100 feet deep. Mudslides and sinkholes, along with additional damage to local roads and highways, have created concern for residents regarding future disaster evacuations. This series of events illustrate the systemic nature of disaster, and it is not possible to talk about them as singular events.

The Fiscal Nature

Municipal and national fiscal policies undoubtedly have an impact on disaster and vice versa. Governments seek to grow their economies through taxation and spending. However, a natural disaster has the potential to undermine that expansion. In crass financial terms, disaster is a liability. As such, when catastrophe hits, governments must cope with the resulting costs without increasing public debt. In our era of global neoliberalism, financial markets are a popular means of dealing with risk. For instance, the "how-to guide" of the International Monetary Fund (IMF) on managing the cost of natural disaster notes the use of insurance, reinsurance, debt instruments (securitization), catastrophe bonds, and other capital market options to transfer risk. The goal, according to the IMF, is to create fiscal buffers against disaster-related financial threats.[22]

Beyond insurance covering potential and actual losses, we see that financialization is central to understanding contemporary disaster capitalism. Finance, simply put, influences everything. Importantly, Sarah Quinn reminds us that this means, "Finance is always social."[23] Financial systems, according to Quinn, not only distribute profit and risk throughout society but are also based on social understandings and relationships that have deepened throughout the last fifty years. Lin and Neely have also argued that "finance has become an essential fabric of contemporary American life."[24] We now see a reliance on market-based solutions for everything. Disaster risk is no different. Importantly, financialization is not just an American phenomenon. Globalization facilitates this process. As John Bellamy Foster argues, financialization, or more specifically, the financialization of accumulation, is at the center of the global economic system. Like other forms of capitalism, this system is predicated on the unequal relationships within countries, such as those between newly industrialized countries in the Global South and wealthy countries that control capital markets.[25] Similarly, this imbalance can be extended to relationships between cities.

What does financialization mean for the fiscal nature of cities and disaster risk? According to Manuel Aalbers, "Urban policy has become a financial instrument."[26] Amid debates over climate change, the mayors of New York, Paris, and Rio de Janeiro penned a 2016 editorial arguing that environmental action begins with cities. Noting that cities "account for 85% of global GDP, and they are engines of technological process and policy innovation," they claim that cities are better positioned to solve environmental problems than national governments. However, in their editorial, the mayors wrote that they have faith in the benefits of expansion and urbanization. In fact, they argue that urban innovation can "increase economic activity."[27] The mayors, in other words, accept an expansion-oriented understanding of cities even as they support initiatives such as renewable energy.

Similarly, following the 2015 United Nations Climate Change Conference in Paris (COP15), hundreds of investors from around the world mobilized around the call for new green technology and energy.[28] This was spurred by Article 2 of the Climate Agreement, which calls for "making finance flows consistent with a pathway towards low greenhouse gas emissions and climate-resilient development."[29] An immediate outcome of COP15 was the convening of the 2018 Ceres Investor Summit on Climate Risk at the United Nations Headquarters in New York. Over 450 financial companies and capital market leaders met to discuss disaster-resilience projects and ways to fund green and sustainable projects that reduce carbon emissions.[30] Meetings such as these reflect a trend toward so-called corporate environmental, social, and governance (ESG) investment practices, which are increasingly influencing environmental policy and action.

While such developments suggest a growing concern for environmental problems, they are deeply rooted in the expansion of profit-making activities that would likely encourage capitalistic urban expansion. ESG investors, while slowly moving money toward renewable energy sources, have not pushed to radically decarbonize our economy. For instance, investment in smart cities still fuels the urban growth machine, which consists of a coalition of policymakers, corporations, and citizens who are true believers in endless expansion. This is not just about getting bigger; it is also about having a wider reach. Expansion-oriented growth means increased local and global consumption. Allan Schnaiberg calls this insatiable hunger the treadmill of production. The treadmill is an ever-accelerating machine that thrives on the consumption of resources.[31] This theoretical approach is addressed in more detail in chapter 3. However, it is essential to note here that our cities run faster than they ever have—by extracting and consuming resources, producing waste, and, as a corollary, creating more risk. As such, we need to question whether our system of risky cities should or *can* continue to run faster and faster.

Why Risky Cities?

Risky Cities is a synthesis of arguments I have previously made that suggest that disaster capitalism is embedded in modern cities.[32] My goal here is to provide a theoretical framework for examining urban and environmental catastrophe. This is also an attempt to ground conversations regarding climate change and disaster in a specific process: urbanization. Cities live with the threat of various hazards, such as infrastructure breakdowns, fire, and sinking foundations. This is not new. However, disaster capitalism has commodified environmental *bads*, creating new threats to urban life.

This project emerged out of my personal and academic interest in wildfire and urban development. William Cronon's *Uncommon Ground* begins with his account of the Laguna Canyon Fire, a conflagration I witnessed firsthand while a teenager growing up in Irvine, California. Wildfire was a seasonal event that you would see on television and smell in the smoky air. The physical transformation of the region exacerbated wildfire risk. I saw urban development creep into Southern California's flammable foothills. As new homes were built over former farms, eucalyptus trees that served as windbreaks were cut down. Orange groves were replaced by brand-new shopping centers, parks, and all the things that come with suburban sprawl. Cronon suggested that "the replacement of nature by self-conscious artifice is an important defining quality of the modern landscape. Irvine is a near-perfect example of the genre."[33] However, nature never truly goes away. At times, it comes back with a vengeance. This book is about how cities deal with—and commodify—this blowback.

In addition to wildfire, *Risky Cities* deals with how different physical *bads*—such as waste and sinkholes—are commodified into financial *goods*. Human settlements continue to grow despite the presence of dangers such as sinkholes, garbage, and fire. Cities sprawl outward and expand into precarious landscapes. While myopic, profit-driven development creates precarity, contemporary capitalism has found a way to integrate these threats into its business models. Urban disaster risk—a so-called *bad* from this perspective—becomes profitable, a so-called *good* that has economic value. Risk means the sale of new insurance policies. The use of government-subsidized insurance—such as in flood- or fire-prone regions—encourages development. Destruction means rebuilding. The existence of such threats become a means to sell new disaster-resistant technologies. However, these forms of risk management do not tackle sprawl, increased consumption of fossil fuels, the production of waste, and other human activities that create risk. Thus, cities steadily become riskier.

Methodologically, this book analyzes primary and secondary historical documents, corporate press releases, white papers, and government data. This is a top-down approach or, more specifically, a critique of those at the top. The social actors in this book are urban policymakers and corporations operating within a global network of cities.[34] These players aim to maximize expansion and profit in a world in which risk is increasing. Put another way, policymakers and corporations seek to maintain a status quo of expansion in a world in which shocks from disaster are looming. At the same time, they have accepted financialization as a solution to these threats. This form of disaster capitalism has become hegemonic. It influences urban policy, public agencies, and private entities such as corporations involved in city-building and insurance firms.

While this book deals with disaster, it is primarily about urbanism under modern capitalism. This focus distinguishes it from other work on urban disaster in several ways.[35] Traditionally, social-scientific research on natural hazards and disasters has examined the social and institutional consequences of catastrophic events. Correspondingly, social scientists emphasize the disproportionate impact of vulnerable places and populations. People who are particularly susceptible to natural and environmental disasters are typically already suffering before the disaster "event." Critically, it could be said that vulnerable populations would remain vulnerable even in the absence of disaster events or environmental hazards. Disasters and their local responses have gendered implications. Men, for example, are more likely to be killed defending their homes while women and children are more likely to die sheltering in place or during evacuations.[36] In the case of evacuation, women and children report higher rates of trauma and stress.[37] Thus, we cannot ignore the social variables that influence the experience of disaster.

Correspondingly, social scientists also examine social and political practices such as the role of government policies that transform climatic, geological, or human events into "disasters." In other words, our environmental context is the by-product of political and economic decisions. *Risky Cities* takes this political-economic approach. Disasters are products of the social and material reality that precedes catastrophe. For instance, powerful institutions such as corporations shape the environment in which catastrophes occurs, as well as a how we respond to them. Social vulnerability, therefore, operates within the context of aggressive urbanization and economic expansion. The push to outsource and financialize everything has weakened society's ability to cope with contemporary hazards.[38] In turn, the lenses of urban political-economy and political-ecology offer a strong critique of neoliberalism's drive toward free-market capitalism.

Plan of the Book

Risky Cities is organized as follows. The first chapter considers how cities live with disaster and capitalism. As mentioned previously, cities have always lived with environmental hazards. What is *new* in our stage of modernity is that cities grapple with risk by integrating it into various economic calculations and institutionalizing it. Chapter 1 expands on what other scholars have called disaster capitalism. Here, I argue that contemporary disaster capitalism is not merely post-disaster rebuilding and profiteering. Instead, capitalists (especially those in cities) have found ways to benefit from the existence of risk. This can be seen in so-called innovations in the built environment that commodify risk. This is also visible in the financialization of nature and the creation of catastrophe bonds, green bonds, resilience bonds, and weather futures—in other words, turning climate data and risk into something like a stock that can be traded in the marketplace. An abstraction or representation that stands in for the climate, nature, and the potential for disaster is created. This is not just about large companies generating a profit from human suffering, which does indeed occur. Rather, urban risk, climate, disaster, city-building, and rebuilding are now tied to financial markets and a variety of commodity exchanges.

This broad argument is supported by examining urban disasters that illustrate the relationship between precarity and city life. The second chapter discusses urban sinkholes. Sinkholes are an example of subsidence or sinking that occur when water drainage problems arise because of human intervention in the landscape. Construction, industrial runoff, cracked pavement, and damaged plumbing are all possible causes of sinkholes in areas with karst or carbonate bedrock. Sinkhole risk is heighted in places scorched by wildfires caused by melted pipes, creating voids underground.[39] Storm runoff, facilitated by human-made impermeable surfaces such as roofs and paved areas, can also lead to sinkholes. The reality of sinking cities reveals that the physical foundations of cities are not as stable as they could be. While proper water management, planning to divert runoff, and the maintenance of infrastructure can reduce the risk of sinkholes, we see that urban expansion and climate change are barriers to the building of more resilient cities.

Sinkholes are also interesting because, in American English, the term is often used to describe wasteful government spending or bad fiscal policy. Money, it is implied, is sunk into an investment that is not profitable. Over the past generation, neoliberalism has driven public agencies away from responsibilities such as disaster relief, environmental regulation, and the provision of other public goods. However, actors such as municipal governments and national governments are necessary for dealing with environmental hazards because of the shared footprint of disaster. This contradiction, as we shall see, produces risk—especially in rapidly urbanizing places.

The third chapter examines the threat of waste. Garbage may appear to be a strange choice for a book on urban disaster. However, as Waqas Butt argues, infrastructure development can serve as a means of "distributing destruction."[40] Municipal solid waste systems move materials that pose a risk to sites that *manage* that hazard. Even functional waste management systems can create new threats, such as pollution and health hazards. This is because risk does not go away; it is simply relocated. We see this in the use of landfills in the periphery of cities or the export of trash and hazardous materials overseas. Another risk is the breakdown of the metropolitan waste collection and disposal system. The absence of an effective waste management system would lead to a shutdown of a city, as seen in cities where sanitation workers have gone on strike or when natural disaster has struck. Moreover, when catastrophe strikes, a great deal of debris, garbage, and other material is created, putting a strain on existing systems.

The generation of waste is not driven solely by population increases and sprawl. Instead, it is also driven by the consumption and growth-oriented ideology of capitalism. Ultimately, growth creates a contradiction in which increased consumption leads to more waste. This waste threatens the viability of existing economic activities. As cities grow amid globalization, they increasingly rely on firms of ever-increasing size for infrastructure projects, as well as other cities buying waste to scale up waste management services. Cities, in other words, are not too different from nation-states that seek overseas expansion. However, this obsession with growth creates a shared footprint of risk and vulnerability for cities.

The fourth chapter looks at the urban political ecology of wildfire. Urban political ecology examines how political-economic processes produce spatial inequalities and environmental injustice. Here, I examine wildfire events to explore how urban expansion and so-called natural and environmental disasters are connected. This chapter examines the relationship between urban development and wildfires (also called bushfires). While political ecologists have studied hurricanes and floods, fire (until quite recently) has received far less attention. What makes wildfires interesting is that most conflagrations are ignited by human activity—camping, cigarettes, arson, trash burning, or downed powerlines. This makes wildfire a problem that is driven by human settlement patterns and exacerbated by climate change.

The fifth chapter investigates policies surrounding mitigation against disaster and building resilient cities. There are debates regarding the usefulness of the concepts of adaptation, mitigation, and sustainability. However, Biesbroek, Swart, and van der Knaap contend that the dichotomy between mitigation and adaptation is overstated. Instead, they argue a mix of short-term and long-term strategies is necessary for adequately responding to disaster-inducing climate change. Specifically, they make the case that the convergence of all approaches lies in emphasizing the spatial aspect of the problem.[41] In other words, the creation of disaster-resilient cities means dealing with the reality that cities—and therefore their residents—are fixed in space and are made vulnerable by their location.

However, what we see is that efforts to reduce risk are often reduced to a series of abstractions. The establishment of international standards for resilience and environmental assessment stands in for safety. While such practices can assist in mitigating catastrophe, they can also be used to justify urban development. While individual building projects are certified as resilient, little is done to address the broader geographic and social dimensions of risk. Again, ongoing urban expansion and consumption are not dealt with.

Risky Cities will conclude with a way to rethink urbanism and disaster. Drawing on Henri Lefebvre's notion of "differential space," I argue that the varied ecological, geographic, and social rhythms of the city can be reclaimed to create cities that are regenerative rather than simple growth machines. Why regenerative cities? While the terms "adaptation," "mitigation," and "resilience" are accepted ideals in contemporary city-building, there are serious concerns that those terms are increasingly co-opted to serve elites rather than those most affected by disaster. Regenerative urbanism means reversing the damage done to society and nature. This requires the cultivation and nurturing of social institutions as well rethinking urban nature by maintaining green spaces and advancing biodiversity. As such, I conclude that an important first step toward regenerative cities will require the de-financialization of risky cities.

1
Living with Disaster and Capitalism

> Someone once said that it is easier to imagine the end of the world than to imagine the end of capitalism. We can now revise that and witness the attempt to imagine capitalism by way of imagining the end of the world.
> —Fredric Jameson, "Future City"

Currently, most of the world's population lives in cities. At the same time, we have seen calamities such as earthquakes, floods, sinkholes, tropical storms, and other catastrophes wreak havoc in urban regions.[1] The existence of these threats means that billions of people live under the threat of natural and environmental disasters (see figure 1.1). This chapter offers a theoretical framework for understanding this precarity by looking at how natural hazards, the built environment, and capitalism contribute to the emergence of *risky cities* in our contemporary world. This framework, in turn, helps us analyze how environmental *bads* that threaten cities can become profitable *goods* that encourage ongoing growth. It also provides insight into why the building of resilient and sustainable cities is not straightforward.

My approach rests on three points. The first is that cities are relatively fixed in space. Indeed, people can move. Builders can shift development from the urban core elsewhere through suburbanization, and companies can relocate

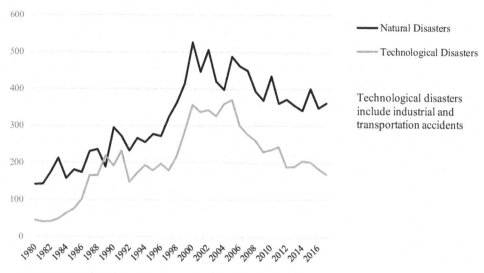

FIG. 1.1 Worldwide major disaster incidents (1980–2017) (SOURCE: The OFDA/CRED International Disaster Database, www.em-dat.net, Université Catholique de Louvain, Brussels, Belgium.)

facilities to other cities. However, this does not happen easily or readily. People have attachments to place. It takes time and resources for people or businesses to relocate. In turn, cities and people are bound by physical and environmental conditions in which disaster (or the threat of it) is a part of their social and spatial fabric. Therefore, urban risk exists because the way cities are situated in a region's geography. This chapter takes this fixity—or the condition of a city being stuck in place—as the starting point for discussing how cities have lived with disaster and how this relationship has changed alongside contemporary capitalism.

The second premise is that the built environment helps commodify natural and environmental disasters. Just as raw material extraction shapes cities, climatic and geologic events also influence urban development. Natural disaster is unique in that it is not commonly seen as an exploitable *good*. For instance, climatic, ecological, and geologic hazards destroy homes, infrastructure, businesses, and factories. Environmental disasters create pollution that disrupts everyday life and kills people. It also, or so it would seem, disrupts the ability of capitalists to generate a profit. City-building, therefore, is a product and consequence of disaster. In turn, we must look at how capitalism—through innovations in the built environment—has figured out a way to monetize the existence of various hazards (see figure 1.2).

Social scientists studying disasters have examined the social and political dimensions of a variety of cataclysmic events—terrorism, chemical spills, hurricanes, floods, and so on. Urban political ecologists and ecological Marxists have examined contradictions within capitalism that undermine sustainability or produce environmental degradation. They argue that environmental disaster is a by-product of capitalism's exploitation of nature. At the same time, this exploitation is central to the process of capital accumulation. This produces a contradiction within capitalism in which this exploitation can eventually become unsustainable and result in a breakdown of the system.[2]

In turn, we cannot understand contemporary natural disasters and the built environment without discussing capitalism. In fact, disaster has a fiscal and physical component. As Kevin Gotham argues, the production of urban space following natural disaster relies heavily on deregulation and financialization. The use of bonds, for example, is an attempt to re-anchor capital and channel money into post-disaster spaces. Bonds are essentially a loan; investors lend money to public agencies or other entities with the promise that they will be paid back with interest. Cities issue bonds to raise money for construction projects or maintenance of infrastructure.[3] The use of bonds suggests that money-making is not just happening through post-disaster profiteering. Here, the production and shaping of the built environment is a means of generating revenue in risky cities. The building and rebuilding of cities involves a great deal of investment. This investment involves many social actors, from construction companies and real-estate developers to investors, insurers, government agencies, and politicians.

This, perhaps, should not be surprising as finance's share of global corporate profits is typically between 40 percent and 50 percent.[4] The influence of this sector leads me to my third point: that contemporary capitalism has transformed environmental bads into goods. Indeed, disaster results in companies profiting from the rebuilding of cities. At the same time, we see that there have been significant developments in the financialization of disaster—such as the creation of resilience bonds and weather futures—as well as investment in start-ups selling new technologies and risk-management strategies. Supposedly, this helps cities manage risk, but it also encourages traditional urban development. As such, this chapter connects the prevalence of catastrophic events, the built environment, and the role of capitalism in commodifying disaster.

To understand this commodification, we can look at the work of Henri Lefebvre. In discussing the notion of "rhythms" in social life, he calls them repetitive, but paradoxically, they appear "natural, spontaneous, with no law other than its unfurling."[5] This describes the rhythm of natural disaster, particularly in cities regularly affected by storms, floods, earthquakes, and other calamities. Disaster is normal yet seems spontaneous. Capitalism, as I discuss later, takes advantage of this rhythm. Capitalist cities—according to Lefebvre—by necessity exploit "biological and physiological experience, nature, childhood, education, pedagogy, and birth."[6] While nature is exploited and damaged by capitalism, according to Lefebvre, capitalism and cities need nature to function.

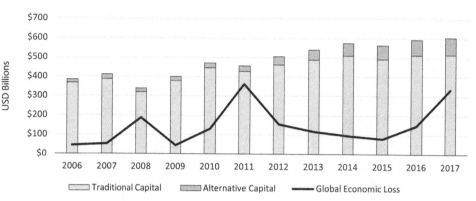

FIG. 1.2 Global reinsurer capital and economic loss (SOURCES: The OFDA/CRED International Disaster Database (2018), www.em-dat.net, Université Catholique de Louvain, Brussels, Belgium; and Aon Benfield (2018), "Reinsurance Market Outlook." Aon plc, accessed December 23, 2018, http://thoughtleadership.aonbenfield.com/Documents/20180103-ab-analytics-rmo-january.pdf.)

Inspired by Lefebvre, my framework is a trialectic consisting of capitalism and the built environment, natural disaster and the built environment, and natural disaster and capitalism (see figure 1.3). In this model, urban disaster risk is a three-part co-constitutive process; in other words, each part is interconnected. This approach is based on Lefebvre's argument that space is multidimensional. He argues that space is made up of the rhythms of daily life, representations and symbols, as well as other social practices. Space, in other words, is not so much a static "thing" as a product of ongoing social relationships and activities. Similarly, we cannot easily disentangle the way the built environment is both a product and consequence of disaster—especially under capitalism. Correspondingly, the building and expansion of cities is not neutral. As Lefebvre writes, the social production of space "is also a means of control, and hence of domination, of power."[7] Who survives and who profits from the building of risky cities maps onto the unequal geographies that reproduce social precarity and political power.

In turn, there are contradictions within this three-part relationship between capitalism, the built environment, and disaster—for example, the development of short-term fixes, which encourages the type of abstraction that disconnects cities from nature. These so-called solutions are found within the urban built environment in buildings and infrastructure. This is also a contradiction, as long-term solutions are ignored and new physical threats are created. Capital adjusts to this contradiction by identifying new ways to commodify disaster and manage risk. This creates another contradiction. The distribution of risk into the marketplace does not mean that risk goes away, nor do the real human

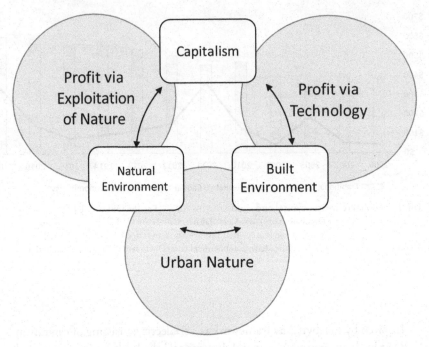

FIG. 1.3 Trialectics of disaster capitalism

consequences of natural and environmental disaster. In turn, disaster can lead to crisis as well as possible openings for resistance and change.

Before proceeding further, I want to emphasize that I am examining so-called disaster and its relationship with cities. All disasters, as well as natural disasters, are social disasters. There is no doubt that disaster is a complex problem influenced by social and ecological forces. The social-scientific study of disaster includes technical disasters such as chemical and nuclear explosions, oil spills, and terrorism in addition to geologic and climatic events. In turn, social and political forces influence how humans experience environmental hazards—such that disasters cannot be truly natural. While cataclysmic events are only recognized when they cause human suffering and financial damage, they are also climatically and geologically real phenomena.[8] In turn, it is critical to look at how ecological, geological, and climatic cycles are intertwined with human rhythms. Cities, after all, have always lived with natural disasters. However, powerful actors in cities have not always profited from disaster or the risk of it.

Returning to my earlier point, cities are relatively fixed in a geographic location. Indeed, boundaries and people move. However, cities and their citizens cannot simply dodge the onslaught of hurricanes, floods, heatwaves, or tremors beneath them. Thus, the fixity of the urban built environment creates risk given its placement. In turn, the existing built environment and the presence of hazards shape the trajectory of cities as they continue to build, expand, and reconfigure themselves. Cities are forced to deal with the anticipation of potential harm and its aftermath because they are stuck in place.

Disaster and the Built Environment: Examining Disaster in Urban Nature

In our technocratic age, urban development and planning fall victim to anthropocentric hubris—or human-centered overconfidence—and the influence of neoliberalism. It is believed that humans are a unique species that can solve any problem and that market-oriented technological fixes are the solution. This anthropocentrism, however, is not new. Richard Sennett suggested that in Western European Protestant traditions, "wilderness" represents "nowhere."[9] Alternatively, this has meant that city-building is framed as being distinct from nature, as humans create "place" out of nowhere. It is precisely this belief system that has made natural hazards social disasters. This anthropocentrism makes it so that we only pay attention when disasters hit *somewhere that people live* and ignore the underlying social and environmental problems that threaten human lives.

The term "urban nature" has meant different things in different fields. For some, it refers to nature within the city. For others, it refers to the metabolism of cities—or the socio-ecological process of urbanization. Urban nature also refers to the life and rhythms of the city. Here, I adopt a metabolic perspective and draw on the work of urban political ecologists who argue cities are hybrid spaces, not unlike a cyborg.[10] This is different from how earlier social scientists viewed the relationship between city and nature. For instance, Louis Wirth in the 1930s famously described "urbanism as a way of life," drawing our attention to forces within cities as influencers of social problems. Wirth claimed that "nowhere has mankind been farther removed from organic nature than under the conditions of life characteristic of great cities."[11] This statement suggests that cities are the antithesis of nature and that urbanization is distinct from ecological processes.

However, this view of a separation between "mankind" and "organic nature" has increasingly been challenged by scholars in environmental history, geography, political ecology, and environmental sociology.[12] Cities are not the antithesis of nature; instead, they interact with and are shaped by nature. Alternatively, Homer Hoyt observed this in the 1930s, noting that bodies of water and other geographic features affect where you can and cannot build as well as how you build.[13] Similarly, hillsides, forests, and waterways are shaped by human activities—such as mining, producing lumber, and irrigation. This also affects where and how people live.

The physical structure of cities is a reaction to environmental conditions, and thus, not only do cities influence nature but they are influenced by nature as well. Anne Spirn, for example, notes the interplay between urban nature and human design. She points to a long history of building on erosion-resistant stone as an example of this relationship.[14] Water, after all, is both the lifeblood of major cities and something that poses engineering and environmental challenges.[15] In many cities, coastal barriers, dams, and levees are shaped by the threat of water hazards. Deflecting storm surges, redirecting water supplies to urban communities, and processing wastewater are all necessities in building cities. However, this can also create new dangers. As we shall see in chapter 2, poorly maintained water systems can cause subsidence and sinkholes.

If city-building is a reaction to geography, then cities built on coastlines, near volcanos, or on mountains are not by coincidence. Bodies of water are used for transportation and food. Volcanoes tend to be surrounded by fertile soil and other resources. Settlements in the wildland–urban interface (WUI) benefit from natural resources such as lumber, as well as scenic landscapes and other amenities associated with nature. Correspondingly, cities prone to earthquake and fire are also shaped by past disasters through rebuilding, the establishment of building codes, and the displacement of communities. This shapes the urban built environment, while the act of building cities changes the environment.

Urbanization under capitalism, it could be said, is a technical strategy to maximize the use of natural resources. The development of architecture, construction technologies, and engineering in the modern world allows the urban machine to be an engine for capitalist growth. Workplaces (farms or factories) and homes are not just spaces that people occupy; they are tools that societies use to reproduce themselves. Infrastructure is not only about the movement of goods and people; it transforms the landscape into a commodifiable resource. The built environment is a means of mastering those resources and the landscape around them to maximize economic production.

When we speak of the built environment, we are not just talking about buildings. Instead, we are examining all the physical parts of the city—streets, infrastructure, and landscape—that are vulnerable to disaster. Take, for example, wildfires in California. Reyner Banham described Los Angeles as consisting of four ecologies—beaches, freeways, flatlands, and foothills.[16] It is precisely this *whole* ecosystem—natural and human-made—that increases fire risk. Hillside communities with coastal plant life connected by highways are especially vulnerable, as human action can ignite the flammable chaparral.

Again, growth and risk are tied to a city's fixity. Indeed, anthropogenic climate change intensifies storms, and hydraulic fracturing leads to earthquakes. However, disasters are also tied to a city's geography. Risk, in other words, is bound to a city's location.[17] Climatic and geologic events are objectively real phenomena that happen in specific places. Such events cannot be grasped without some acceptance of traditional notions of the urban or natural disaster as an objectively real phenomenon. At the same time, we cannot fully understand disaster within urban nature without looking—in new ways—at the built environment as a construct that interacts and copes with catastrophes.

Take, for instance, trash. Before the twentieth century, solid waste was typically dumped in nearby waterways, burned, or buried. Even with the human-made crises of industrial capitalism, cities were and are limited by their physical location. Cluttered streets and polluted rivers limited urban growth and expansion. London experienced the "Great Stink" in 1858 because of waste on the banks of the Thames River. Early efforts to deal with waste, such as incineration, often created new problems. In the early twentieth century, trade publications such as *Fire and Water Engineering* published articles calling for education campaigns discouraging trash burning near buildings. Later, there were calls to ban backyard waste incineration because of fire risk and air pollution.[18] Over time, regulation moved waste and the risk of fire from incineration further and further away from buildings.

However, the risks associated with waste did not go away. Eventually, waste was moved into the urban hinterlands for processing—such as New York's "piggeries." While municipal solid waste was pushed outside city limits, growing cities crept closer and closer to these wastelands and, in some cases, expanded *over* landfills.[19] Waste thus became part of the built environment. Cities are often built on the rubble and landfills of the past. For instance, in New York City, the Collect Pond was so polluted that it was filled in during the early 1800s, and structures such as a jail were built on the site.

As late as the 1980s, waste products from factories were used as "clean fill"— that is, material used to fill in construction sites or change the grade of a development. This would, at times, result in asbestos, lead, or other toxic materials later being found in residential communities.[20] In 2016, it was discovered that a developer in Woodstock, Georgia, had built a neighborhood on buried construction debris and other material. Over time, the waste gave way and created a sinkhole.[21] Even today, construction sites use fly ash (also called bottom ash)—a waste product from the burning of material—as fill or a construction product (see chapter 3).

These examples suggest that urban nature is precarious and full of risk. The built environment is often literally on shaky foundations—either because it was built on top of waste or because of other hazards. Despite this, people cannot simply get up and move away, and cities are slow to rework their infrastructure before disaster strikes. Still, contemporary cities grapple with risk by integrating it into various economic calculations. Risk management is now institutionalized and embedded in urban governance structures.[22] Undoubtedly, failures in urban planning and policy create hazards. Profiting off risk, according to Naomi Klein, is now part of a disaster-industrial complex.[23] This is different from just post-disaster profiteering, and not unlike the change in warfare with the advent of the military-industrial complex. The post–Cold War military-industrial complex is different from just selling weapons for profit, just as the disaster-industrial complex is different. While blacksmiths have profited from war in the past, just as carpenters benefited from rebuilding homes, the complex corporate interlocks and institutions that allow capitalists to profit from war or disaster were absent in the past. The military-industrial complex now consists of policymakers, defense contractors, and think tanks that benefit from war making—including consulting, and logistics services—not just selling weapons. Similarly, the financialization of disaster risk is different from post-disaster rebuilding. Rather, we see banks, financial services, insurance, real estate, banks, and other industries such as construction, emergency services, energy, and waste management developing new ways to commodify pre- and post-disaster risk.

In other words, the aforementioned actors are now operating within an environment where disaster risk connects them to the stock market, exchanges, and other institutions that handle securities or tradable financial assets. Carruthers and Kim note that finance, from a sociological standpoint, often involves nested relationships that provide the context for action.[24] Insurance, however, plays a particularly prominent role in shaping this context. In turn, Kate Booth calls insurance an example of financialized "worry." These anxieties, however, are spread throughout everyday life. Everyday worries, such as fear of a disaster destroying one's home, are embedded in insurance policies governed by algorithms that reduce risk for investors all over the world.[25] Insurance, therefore, links households and whole neighborhoods to international financial institutions.

Capitalism and the Built Environment: The Built Environment as Technology

This set of nested relationships and worries are part of the disaster-industrial complex. This context is one under which adaptation, mitigation, and resilience projects—despite their intent—operate. While urban nature and social life should not be reduced to commodification and capitalist relations, we cannot ignore that the built environment in the contemporary world is a tool of capitalism. Urbanization is a technical strategy, and cities are machines configured to maximize the use of natural resources. However, not all aspects of nature are easily exploitable environmental goods such as agriculture and natural resources. New technology and profit-making strategies are needed to transform environmental *bads* into *goods*. Investments in science and technology, as well as anti-disaster social expenditure, prevent the destruction of productive fixed assets such as homes, stores, and other aspects of the built environment. In turn, the development of disaster-resistant architecture, coastal armoring, and other technologies often serve the interests of capitalists.

Here, the field of science and technology studies is useful. The built environment is a technical intervention that is undoubtedly necessary for the construction of modern cities. However, technology and science cannot be separated from the social, economic, and political. Technocratic decision making employed by policymakers has intensified alongside the development of capitalism. As Karl Marx points out, in a capitalist society, "the conscious technical application of science" grows in scale.[26] Marx's critique of science is further developed by the Frankfurt School and other scholars who have argued that so-called value-free science is often an instrument of capitalism.[27] Technology, after all, has played a central role in alienating workers from their labor in the name of efficiency.

Furthermore, technology has furthered contemporary capitalism's tendency to abstract everyday life, social relations, and nature. Not only does technology dis-embed social relations, but the built environment becomes disconnected from nature. This abstraction facilitates the financialization of everything. Take, for instance, the rise of Big Data and algorithms in dealing with disaster. Contemporary risk management draws on data to generate models of possible events, measure the level of threat or hazard, and assess the likely vulnerability of things such as buildings, all of which is translated into potential financial loss. Catastrophe modeling, it is argued, can help planners mitigate damage from potential disaster events. However, Lefebvre warned us that once quantifiers of reality reach their goal, "then suddenly concrete reality vanishes."[28] Not only do we get formulas standing in for nature, but we have stand-ins for safety. The concrete reality of safe spaces is replaced with a check mark on an assessment report. These stand-ins allow for weak mitigation plans and permit ongoing expansion.

This estrangement has intensified in contemporary capitalism. Big Data has widened the gulf between abstractions and reality. We see this most clearly in technical products derived from algorithms that yield ever-increasing profit for elites. Cathy O'Neil, noting the rise of Big Data and powerful computers, has questioned the use of algorithms. While the predictive models have the potential for good, she argues that algorithms make for an opaque society and threaten democracy. O'Neil raises the concern that most people—including those who implement algorithms in decisionmaking—do not know or understand how such models work. It is a so-called black box that lacks transparency.[29]

Of concern is the growing role of such technologies in policymaking. City-building and disaster management increasingly rely on Big Data, as well as tightly controlled proprietary algorithms, that technocrats use to legitimize policies. We need only look at the rise of "smart city" initiatives that attempt to optimize day-to-day operations in cities through data collection and new technology. Not coincidentally, this coincides with the use of scientific advisory committees in public policy. Sheila Jasanoff argues that instead of problem solving, committees made up of technocrats instead create policies that serve elites and capitalism's search for profit.[30] Recently, this has meant a push for artificial intelligence, increasingly complex modeling, and other commercial technologies. Indeed, there are concerns and criticisms within the insurance industry and other sectors that an emphasis must be placed on better quality data rather than blind acceptance of the validity of catastrophe models.[31] However, despite these concerns, it clear that this technocratic rationality has won out, and it now shapes contemporary city-building.

David Harvey's analysis of how capital flows through different circuits can also help us further understand this relationship between the built environment and capitalism. Capital, or financial wealth, flows through different circuits. The primary circuit is invested is commodity production—that is, making things. Sometimes, this leads to overproduction—or too much of a commodity on the market—leading to falling rates of profit or crisis. To deal with crises of overproduction, capital can be rechanneled into a secondary circuit. This involves investment in fixed assets such as upgraded factories and spaces for consumption such as stores, homes, and infrastructure. According to Harvey, the built environment and urbanization is an essential process in the functioning of capitalism. Circuit switching not only prevents a crisis tied to overaccumulation but allows for consumption and other ways to generate profit. Finally, if there is overaccumulation in the secondary circuit—for example, too much construction—then capital can be channeled into the tertiary circuit of science, research, and design. Here, technology is used to enhance labor productivity as well as create new means of consumption. Expanding on Harvey, I argue that disaster encourages circuit switching.[32]

Technologies in the built environment that battle disaster—coastal armoring, floodgates, and seawalls—obscure the relationship between city and nature, as well as allow for the sinking of capital into the built environment. A multimillion-dollar seawall project, for example, gives residents, insurance companies, and policymakers the illusion of safety when it is in reality a temporary fix. In the meantime, it allows cities to expand across risky coastlines, generating wealth for developers. However, this has the potential for increasing disaster risk, thus driving more investment, production, and consumption of technology.

The deployment of innovative new architecture, engineering solutions, and other strategies or technologies is not merely a form of profiteering. It is a means of dealing with vulnerable infrastructure while generating new forms of profit. Cities go through cycles of building and rebuilding following disaster. At the same time, it is how the built environment—as a fixed asset—is made useful again and produces more value. This applies not only to post-disaster neighborhoods that were damaged but also to undamaged retrofitted older homes that are now more disaster resistant. Contemporary urbanization, therefore, is not merely tied to population growth and need. It is a tool for increasing profit margins and the rate of financial returns.

Typically, policymakers do not question whether ongoing expansion is safe or sustainable. Recovery itself is already complicated. Even more problematic is whether rebuilding itself is safe. In turn, disaster-resistant (but not disaster-proof) homes serve as a placeholder for safety, giving property owners and buyers, as well as insurers, a sense of security—which may prop up property values. This emphasis on an engineered solution is reflective of a technocratic rationality that encourages urban growth and development as well as renewal projects through so-called innovation. Indeed, disaster alone does not drive innovation. Disaster is just one of many crises—for example, fiscal crises—that can encourage circuit switching.

While the tension between city and nature has long existed, it is only in our contemporary stage of capitalism that disaster risk has become a marketable algorithm. Nature and disaster now consist of variables that can be controlled for. Models of weather patterns, disaster risk, and the cost of rebuilding are proprietary formulas that are sold and exchanged. Indeed, well-implemented technology can save buildings and lives. Forecasting for weather events and sensors that anticipate geologic catastrophe are useful. Embracing such technology can mitigate or reduce potential harm. Yet, relying solely on such systems ignores more complex questions of planning, land use, and urban nature. The result is short-term technofixes, whereby capitalism negates the potential of building truly safe cities.

Climatic and geologic events are natural, yet humans do seek mastery over them.[33] As Lewis Mumford argued, "In the act of 'conquering nature,' our ancestors too often treated the earth as contemptuously and as brutally as they treated its original inhabitants"[34] This contempt for nature creates new risks. The technocratic rationality of neoliberal policymakers pushes them to try to commodify environmental *bads*. This results in hazards that are mostly blowback for questionable policy. Blowback typically refers to the unintended consequences of military intervention. The attempt to quickly mitigate environmental problems while generating revenue from the solutions is not unlike military interventions such as selling arms to solve immediate foreign policy problems leading to longer-term regional political destabilization.[35] Even the most well-intentioned pre- and post-disaster plans can have unintended impacts on those affected. As Frederick Engels suggested, we should not "flatter ourselves overmuch on account of our human victories over nature. For each such victory, nature takes its revenge on us."[36]

Growth-oriented neoliberal policies are commonly green-washed as to hide their unsustainability. JoAnn Carmin, for instance, notes that funding for disaster mitigation and relief is dwarfed in comparison to investment in urbanization.[37] As we shall see in chapter 3, waste-to-energy or garbage incineration is sold by multinational corporations as being "green" since it removes the trash. However, incineration—as well as recycling—does little to curtail carbon-generating consumption that creates new crises.[38] Just as there are concerns about "hijacking sustainability," we must be concerned that disaster mitigation and resilience has been appropriated by those who benefit from catastrophe.[39] Mitigation technologies are used to sell safety to developers, businesses, and citizens allows the disaster-industrial complex to take advantage of cyclical disasters. This, however, is not a long-term strategy to make cities less risky.

Disaster Capitalism: Financialization and Creative Destruction

Post-disaster, there is no doubt that capitalists eagerly invade a city—seeking real estate development, construction, rehabilitation, and the marketing of new technologies. That said, experts and entrepreneurs are aware which pre-disaster cities and regions are prone to catastrophe. This awareness has done little to stop precarious developments, as risk management becomes increasingly profitable. As such, we must distinguish between post-event profiteering of the past and what is new in the disaster-industrial complex.

Capitalism's relationship with nature is well studied—especially by Marxian theorists. Notably, this literature has a strong emphasis on crisis and contradiction. James O'Connor argues that capitalism increasingly commodifies the environment, creating a contradiction—one where capitalists now must spend more and more to sustain the system that they are destroying.[40] Similarly, John Bellamy Foster argues that the exploitation of nature will result in a metabolic rift or ecological crisis. Modern society has a metabolism that involves an ever-growing consumption of goods and the production of waste. Foster contends that capitalist agriculture has the potential for undermining both the fertility of nature and society's ability to reproduce itself as nature (e.g., soil) is devalued and the conditions for production are exhausted. Rather than let fields fallow, more fertilizer is used to sustain nutrient levels in soil. This requires more intrusive human action with an ever-increasing potential for ecological disaster.[41] Taking a slightly different approach, Jason Moore presents the notion of capitalism-in-nature, whereby the modern world is dependent on cheap food, energy, or natural resources. However, the mechanism under which these resources are extracted and used are increasingly threatened by environmental degradation. Pollution, climate change, and ecological disasters can undermine the ability of capital to generate new forms of income and result in falling rates of profit and thus crisis.[42]

This literature for the most part emphasizes environmental *goods* to be exploited by capitalism and how this commodification of nature results in crisis. However, what is important, in the case of disaster capitalism, is how capital ends up offsetting damage by spending more and more on coping with that risk. We must remember that calamity—until recently—has not been a systemic generator of profit for capital. By systemic, I mean that disaster or the threat of it generates profit across multiple sectors. Previously, revenue generated came from post-disaster profiteering. Indeed, profiteering is still a part of disaster capitalism. There is ample evidence of the predatory nature of capitalism following natural disasters.[43] However, this profiteering is now just a part of an integrated risk-management industry.

Disaster industries have figured out ways to move beyond post-disaster events but also encourage profitable management of risk.[44] There are two ways disaster is transformed into an environmental good. The first is turning disaster events into acts of creative destruction. The second is the financialization of disaster and risk. Historically, this meant providing insurance. Insurance, however, only works if growth and surplus revenue is being generated. Marx reminds us, there must be surplus profits beyond the replacement of things destroyed by disaster. As insurance must be made from surplus profits, this means that there must be continuous overproduction.[45] In turn, insurance is dependent on endless expansion. While insuring productive fixed assets in the built environment offsets financial risks, capital is always searching for new sources of profit. This could mean the construction and sales of more buildings and the issuance of more insurance policies. Again, this necessitates overproduction. Yet there are impediments to physical growth. There are geographic and spatial limitations to sprawl—for example, existing structures impede development—and resistance to urban renewal.

This means the built environment is fixed in space, which can limit economic expansion. However, capitalism has managed to deal with this problem in several ways. Amid an economic downturn, capital resolves this fixity problem through so-called creative destruction. Crisis, as Joseph Schumpeter famously suggested, can lead to "creative destruction." That is, destruction can spur innovation and the development of new means of wealth generation. The burst of a financial bubble, for instance, may force investors to look elsewhere for new sources of revenue. Similarly, the bursting of a dam creates new opportunities for investors. This is echoed by Lewis Mumford suggesting that "from the standpoint of an expanding capitalist economy, indeed, capitalism's prospects of profits, which rested on continuous turnover, demanded the continued destruction of old urban structures, for the sake of their profitable replacement."[46] In the absence of an economic downturn, disaster can serve the same role as a financial crisis. Richard Hornbeck and Daniel Keniston argue that the Great Boston Fire of 1872 served as creative destruction—encouraging the replacement of outdated buildings and fostering expanded economic gains.[47] Put another way, when disaster occurs, this can be a critical juncture for both capitalists seeking development in otherwise inaccessible spaces. Neighborhoods that could not be redeveloped are now ripe for development because of residents being forced out by disaster.[48]

Similarly, Kevin Gotham and Miriam Greenberg have examined the role of catastrophe—or, more specifically, cycles of crisis—that have driven urban and economic development. They find that crises ranging from 9/11 to Hurricane Katrina can serve real estate interests.[49] New York and New Orleans have seen a great deal of post-disaster rebuilding and investment. We see this elsewhere around the world as well. After the 2010 Haitian earthquake, an industrial park and power plant was planned on the site of the fishing village of Caracol. The project was funded by the U.S. Department of State, the Clinton Foundation, loans from the Inter-American Development Bank, and a Korean clothing company set to be a tenant at the park. Here, disaster-generated crisis directly led to a dramatic transformation the built environment.[50]

While I use the term "creative destruction," capitalists cannot simply intend for natural disasters to occur. However, they can anticipate consequences and outcomes when disasters do occur. Alternatively, capitalists can know that a disaster *will* occur but not know *when*. Capitalists clearly know what they want after a disaster. Policymakers in post-disaster cities typically open the door to neoliberal reform because of the immediate need for rebuilding. In addition to the issuance of contracts for builders to repair the city, they recruit think tanks and other interests to intervene. This becomes more than just rebuilding physical structures. It involves the reconfiguration of public institutions and the built environment around neoliberal ideals—which also sets the stage for how cities deal with future catastrophes.

The approach of these institutions, run by neoliberal policymakers, is increasingly tied to the disaster and risk management sector. Scott Knowles has looked at these real estate interests—as well as other actors involved in handling catastrophe and risk—in his examination of "disaster experts." He suggests that those working in this area often promote and develop "creative adjustments to new risks and management of old ones."[51] Indeed, those working in this area are trying to solve the problem of increased risk. In recent years, this has meant the development of new tradable financial instruments or securities that shift risk into the marketplace as well as new ways to fund mitigation projects.

These solutions, however, are short-term technofixes that operate within the context of contemporary disaster capitalism. Here, catastrophe becomes a form of creative destruction that allows for economic and urban growth. We know that disaster does not deter business interests, including real estate. Mark Skidmore and Hideki Toya found a positive correlation between natural disasters and capital accumulation.[52] Similarly, James Elliott and Matthew Clement find that following a disaster, land development is accelerated.[53] In turn, the social science literature suggests that catastrophe can be a tipping point for real estate, whose actors can either rebuild or conveniently liquefy real estate previously constrained by the presence of residents or legal restrictions on demolition. The strategies deployed reflect a form of rent-seeking behavior. Typically, revenue is generated by real estate in the form of rent, rather than wages or profit. Natural disaster allows for developers to liquefy assets or move investments by building elsewhere.[54] Disaster opens up valuable real estate for development, allowing the sector to generate new forms of rent. For example, in 2005, after Hurricane Katrina, Baton Rouge representative Richard H. Baker infamously told lobbyists: "We finally cleaned up public housing in New Orleans. We could not do it, but God did." The outcome has been gentrification in New Orleans because residents were unable to return and rebuild. In turn, Gotham has argued, the neoliberal state has assisted in the anchoring of capital in the Katrina-devastated Gulf Coast.[55]

Here, we see a broader risk-management industry in which all aspects of the environment, such as extreme weather events, are commodified. It is not just insurance covering extreme events but investors betting on environmental conditions. This is covered in greater detail in chapter 5. However, this is a market of growing importance, whereby investors and financial services companies are buying and selling contracts that transform different weather or disaster scenarios into profitable outcomes. For the disaster-industrial complex, this is rationalized as a means of risk reduction. It goes beyond simple insurance, which largely focuses on losses when unlikely events occur. Temperature fluctuations and less damaging events can have an impact on a variety of sectors. Scholars have noted the increased use and sales of weather derivatives, hurricane futures, and risk models in a variety of different sectors—such as energy companies, insurance, agricultural firms, or even the tourism industry.[56] The benefit of these instruments, according to financiers, is that they shift risk from a single company into the market—or to the many gamblers betting on disaster.

Disaster *management* is no different from other forms of marketization and securitization to generate profit.[57] In the case of natural disaster, prices for catastrophe bonds are based on the prevalence of disaster. If it is a quiet season, prices do not change. This also means that speculative traders are unable to take advantage of price volatility. However, a series of disasters could increase trading in the market. For example, hurricane futures attract speculators betting on disaster. If it is anticipated that a tropical storm will grow into a hurricane before making landfall, a trader who buys a futures contract stands to profit if the storm makes landfall as a category 4 hurricane but could lose the cost of the contract if the storm ends up as a mere tropical depression. Similarly, a trader who doubts the storm will make landfall or maintain its strength could sell their existing futures contract on a category 2 storm and profit.

The rise of disaster-oriented financial instruments would not be possible without a technical infrastructure for calculating risk. Put simply, the commodification of nature requires an obsession with technologies such as sophisticated algorithms. Weather exchanges such as the Chicago Mercantile Exchange (CME) rely on the National Climatic Data Center and EarthSat for data for their indexes. In the case of other disasters, such as sinkholes, the National Aeronautics and Space Administration (NASA) and the European Space Agency (ESA) have been developing radar and satellites for monitoring potential hazards. In addition, companies large and small are increasingly involved in the generation of climate- and disaster-related data. The Raytheon Company, a defense contractor and the world's largest producer of guided missiles, provides weather forecasting for the National Weather Service, the National Oceanic and Atmospheric Administration (NOAA), and companies that may be interested in such services. In other words, Raytheon—a company that specializes in the destruction of cities—also provides the resources to manage disaster risk. Marketing its Advanced Weather Interactive Processing System, the company advertises training modules to build a "weather-ready nation." Raytheon, in this service, offers detailed analysis of weather data from satellites to provide up-to-date models that can be used for disaster planning and recovery.[58]

Interest in climate data is visible beyond corporations like Raytheon. Financiers, start-up companies, and the entrepreneurial ventures that typify Silicon Valley have become interested in climate data as well. NASA and NOAA have begun exploring outsourcing data collection to companies that can launch smaller satellites more quickly than government agencies. In 2018, the market for small satellites (those that weigh under 1,000 kilograms) was projected to be a multibillion-dollar industry, as start-ups try to compete with Elon Musk's SpaceX.[59] While more satellites can lead to more accurate or detailed climate data for scientific research and weather forecasts, the financialization of this data undoubtedly fuels the disaster-industrial complex. It has dramatically expanded how companies can profit from disaster and risk through data collection, modeling, and selling solutions. While coverage of this phenomenon in science and technology publications such as *Wired* is fitting, we now see business magazines such as *Fortune* covering these start-ups and describing them as "booming."[60]

This data is put to work in catastrophe models that the private sector, as well as government agencies, increasingly rely on for urban planning purposes, development, and governance. However, hazard mitigation systems involve many for-profit players. Risk managers rely heavily on CoreLogic's EQECAT, AIR Worldwide, and RiskLink from Risk Management Solution (RMS), which are all proprietary catastrophe risk models. These models emerged in the late 1980s and early 1990s in response to major disasters such as Hurricane Andrew. They have since expanded to include models that quantify terrorism and infectious diseases. Leigh Johnson argues that such modeling plays an essential role in making disaster futures and other insurance-linked securities exchangeable in the marketplace. Like the earlier discussion regarding "black box" algorithms, Johnson likens the producers of models to credit rating agencies—again, obscure formulas that have a profound impact on real people's lives.[61] These proprietary models play an integral role in the underwriting process for various insurance policies—such as those related to property and real estate development. Despite the lack of transparency, they have a profound impact on the marketplace. Insurance, after all, is often needed to cover mortgages.

Like the rise of start-ups that generate the data fueling these models, new start-ups are looking to harness the power of "machine learning" to generate disaster models like a Silicon Valley venture.[62] Start-ups typify contemporary capitalism's high-risk approach to generating investment—which does not necessarily equate to solutions. The rise of start-ups involved in natural disaster forecasting is troubling, as these ventures have high rates of failure. Start-ups are a financial gamble full of busts. The examples above suggest a deepening of what Susan Strange called "casino capitalism"—deregulated markets with boom and bust cycles tied to speculation.[63] At the same time, the boom and bust goes deeper than just financial markets and investors' portfolios. Instead, the risky economic underpinnings of the market increasingly shape how politicians and companies think about cities and city-building.

Despite the well-intentioned development of proprietary systems, the underlying forces that lead to their deployment is the building of risky cities and a focus on technocratic market solutions to such problems. The underlying causes of urban precarity—growth, climate change, and the production of waste— are often unaddressed. The keyword here is "managing" risk. Risk is handled as if it were a stock portfolio—a well-planned gamble to maximize profit. Put another way, the goal—not unlike a Vegas-style casino—is to minimize losses and always come out on top.

Of course, data can also be used effectively to save lives. Following the 2003 heatwave in Europe, there was a massive deployment of heat health warning systems in major cities. The development of such systems included the creation of algorithms that estimate potential mortality based on meteorological conditions, investment in infrastructure, and emergency services.[64] Another example is New York City's Emergency Management department. While it has its origins in World War II and Cold War emergency prevention, natural disasters, extreme weather, and influenza pandemics are now included in its charge. Since Hurricane Sandy in 2012, New York City has steadily increased the role of computer modeling, weather forecasting, and loss estimate models in its hazard mitigation plans.[65]

While proprietary models dominate the market for catastrophe modeling for insurers, several consortiums, partnerships, and other projects provide open-source models for cities. Public agencies, after all, also must manage risk. Central

American countries developed the Central American Probabilistic Risk Assessment (CAPRA) platform in 2008. The product of a partnership among the Center for Coordination of Natural Disaster Prevention in Central America, the United Nations, the Inter-American Development Bank, and the World Bank, CAPRA provides a means to model regional hazards. A goal of this project is to have a nonproprietary, open-source system for risk analysis that allows for collaboration and ongoing application development. This also includes the ability to crowd-source data, which has the potential to be more collaborative and democratic.[66] What it suggests is that it is possible to address urban risk without reliance on profit-driven planning tools and perhaps to build less risky cities.

Blowback and Shared Risk

This chapter offered a trialectic framework consisting of capitalism and the built environment, natural disaster and the built environment, and natural disaster and capitalism. My argument is that this trialectic—a three-part co-constitutive process—helps us understand both the city and the environment when it comes to disaster. While cities are relatively fixed, the consequences of disaster are often not bound to a single place. The displacement of human beings, systemic financial loss, and the breakdown of intercity connections point to the idea that environmental vulnerabilities are shared within regional and international networks of cities. Growth and expansion—crucial parts of capitalist ideology—drives cities to externalize costs as well as risks to other cities. As such, real estate developers potentially benefit as much from creating hazards as they do from mitigating against them.

Capitalism's influence on disaster readiness can be compared to the pharmaceutical industry. It has been noted that vaccinations—the equivalent of creating resilient cities in the human body—is not as profitable as developing drugs that mitigate health problems, for example, anti-cholesterol medications. This is because vaccines prevent illness. They solve the problem. Anti-cholesterol drugs are taken for years, creating a perpetual stream of "customers." The result has been fewer companies producing vaccines and less funding for their development by major corporations.[67]

Disaster management has consequences that cascade or spill over across all of society. Pre-disaster fiscal policy includes whether to use public funds on things such as firefighters, wildfire defense, and early-warning systems, as well as how to raise money for these initiatives amid neoliberal anti-tax sentiments. Post-disaster, there are intense debates over what "recovery" entails. Rebuilding involves repairing damaged structures as well as rebuilding lives, communities, and livelihoods. Yet, who and what is prioritized is not only a political and economic consideration but a decision making process with serious social consequences.

A consequence of policy grounded in disaster capitalism is blowback—an unintended consequence or repercussion. The term was popularized by Chalmers Johnson to describe a form of retaliation for foreign policy and military decisions. Importantly, he notes that "because we live in an increasingly interconnected international system, we are all, in a sense, living in a blowback world."[68] This global system, however, includes an environmental dimension. Andrew Price-Smith reminds us that "we remain deeply embedded within the global ecology."[69] He argues that our global interdependence has a profound impact on our ability to cope with environmental threats, including disease. In turn, decisions such as war can have serious blowback. In the case of urban disaster, the blowback is a by-product of myopia or hubris on the part of expansion-oriented policymakers. For example, growth-oriented urban development policies may be an economic victory locally but may result in social and ecological blowback elsewhere. Not only do such policies ignore the social, political, and economic problems of expansion-oriented growth, but environmental *bads* are externalized. Indeed, in this global ecology—in which cities share an environmental footprint—hazards cannot be reduced to singular, local, or even regional problems.

The shortsightedness of neoliberal policymakers results in urban planning that creates short-term solutions but worsens existing problems such as inequality and climate change. Policymakers justify such strategies as mitigating disaster while avoiding economic hardship. The resulting blowback, however, can come in the form of political scandal (e.g., cronyism in building projects), citizen outrage, and in some cases, economic crisis, increasing new risks and catastrophe. We must remember that there is no such thing as a purely natural disaster: such catastrophes are always social and political.

Importantly, the profits generated by catastrophe do not benefit most citizens. Because of the connection between physical and social vulnerability, those most affected by disasters are those who have the least power in society. A European Investment Bank (EIB) working paper discussing strategic adaptation to climate change asks a fundamental question: "Who should adapt?"[70] The paper notes that adaptation is a public policy issue and of concern for the private sector. In an ideal situation, public and private actors *should* work together to protect those who are vulnerable to disaster. However, the paper also warns that public–private partnerships are fundamentally flawed in that liability is uneven. Specifically, the financial burden usually falls on the government agency, not on the contractor. Unfortunately, this is the typical outcome of neoliberal policies.

The blowback for capitalist growth-oriented economic and urban policy is most felt by those most vulnerable. In general, we see that those facing extreme inequality suffer the most significant losses yet receive limited assistance in recovery efforts. Again, those who are most vulnerable to the physical threat of disaster are also vulnerable to social and economic hazards as well. Who lives where and has access to what kind of resources are already affected by citizenship status, race, class, and gender.[71] For example, immigrants might not receive warnings or evacuation notices in their native language, and they might have trouble getting information on how to access post-disaster aid and resources to recover. In recent California fires, we have seen that immigrant agricultural workers living in the WUI are often right on the fire line. These predominantly Latinx farmers were either not warned or were not provided evacuation orders in Spanish.[72] Clinics and organizations that serve immigrant communities are often already under-resourced and additionally strained by significant wildfires.[73] As such, many suffered smoke inhalation and did not receive adequate medical treatment either during the wildfire or afterward.

This is not limited to wildfire. Not all buildings are necessarily up to code. Not all communities have access to adequate infrastructure. Flawed zoning exacerbates risk during hurricanes. Migrants in developing nations have no choice but to live in vulnerable areas.[74] These are social disasters.[75] In turn, we see that those with more resources—whether it be individuals, communities, cities, or nations—are more likely to have the resources to rebuild, mitigate, or adapt after a catastrophe.[76] Alternatively, we see that lending agencies may help developing countries in recovery but are less involved in the development of mitigation or adaptation plans.[77]

Yet, if existing policy only fuels disaster capitalism, we must think more critically about what really constitutes resilience or adaptation, or whether such frameworks are actually effective. In the ensuing chapters, I illustrate how modern capitalist cities have turned natural disaster—an environmental *bad*—into a commodified *good*. I will offer examples of disaster capitalism in how cities have dealt with sinkholes, garbage, and wildfire. While very different forces drive the creation of these hazards, they are all by-products of urban development and the fixity of cities. Immobile settlements must identify ways to mobilize waste or suffer consequences. While mitigation strategies offer *some* protection to homeowners who can afford it, we must be wary of the way in which contemporary capitalism has structured the relationship between the built environment and nature. Thus, the analysis of future policies and solutions must be understood in relation to social and environmental problems. Simply put, we must think critically about urban nature and its relationship with disaster.[78]

This chapter may seem cynical regarding mitigation strategies and modern cities coping with disaster. Undeniably, we should not simply reduce all city-building and disaster responses to capitalist development. Urban nature and disaster are also influenced by remembrance and forgetting. This is perhaps where a critique of political-economic approaches is valuable. Cities are not just spaces. They consist of places that persist over time. In Walter Firey's classic critique of the Chicago School, he importantly noted that "locational activities are not only economizing agents but may bear sentiments which can significantly influence the locational process."[79] Just as cities inherit spatial patterns from nature, landmarks and open spaces do from human history. The power of such places, which includes rivers, parks, greenbelts, and other areas, makes urban nature more than just a by-product of economic processes.

Ultimately, we must look for a better way to deal with urbanization, given the human-impact of disaster risk. In a perfect world, the solution is to avoid new building in hazardous areas, such as karstic regions, flammable landscapes, and floodplains. In cases where construction is taking place, it means taking into consideration drainage, landscape management, and other elements in the environment that produce hazards. Moreover, new buildings should have a regenerative component that seeks to remediate the damage caused by past urbanization.

2

Sinkholes and the Risky Foundations of Cities

> Our era is trying to eliminate the tragic, while all the while it is sinking deeper and deeper into tragedy.
> —Henri Lefebvre, *Introduction to Modernity*

Sinkholes are phenomena that seem to appear out of nowhere to create visually arresting disasters. At times, they cause devastation, not unlike a cinematic version of an earthquake, as the ground opens to devour homes, cars, and people. In 2013, a massive sinkhole consumed a resort near Walt Disney World in Florida, and in Guangzhou, China, a sinkhole swallowed up buildings and disrupted power to the neighborhood. In 2014, a sinkhole opened under the National Corvette Museum in Kentucky, dragging several cars into it. In 2015, sinkholes swallowed cars and homes in London leading to evacuations. The list of buildings, vehicles, and people swallowed into the earth is extensive. In other words, sinkholes are a common problem. While sinkholes do not receive the attention that is paid to other disasters, their ubiquity reveals a great deal about urban risk.

Urban sinkholes are less a product of the earth's fury, such as an earthquake, and more a product of water, geology, and the built environment. Sinkhole activity is linked to the way humans establish a built environment that exploits natural resources and geography. Generally, sinkholes form in karst terrain—regions underlain by bedrock that can be dissolved by water. Bedrock containing salt, gypsum, anhydrite, limestone, or dolomite is particularly susceptible to dissolution by rainwater that has percolated into the ground; this dissolution creates cavities that grow too large to support their roofs, resulting in collapse. The bedrock dissolution processes that form sinkholes can occur over the course of months or years, but collapse is usually catastrophically sudden.

Several factors affect where bedrock dissolution and consequent sinkhole formation occur. Overabstraction, or the dramatic removal of water from a natural source, can significantly accelerate bedrock dissolution underground by drawing water *through* the rock and consequently promoting the formation of sinkholes. This process is common near mining sites and settlements that rely on local groundwater for drinking or agricultural use. Variations in the permeability of the ground also affect where rainwater can interact with bedrock through cracked pavement and water diversion. Other sources of water, such as damaged plumbing systems, can also promote sinkhole formation. Finally, some sinkholes form by water physically winnowing away earth to create collapsing voids in a process called pseudokarst formation.

Because all these processes occur in the very ground that cities are built on, it can be said that sinkholes reveal the risky foundations on which many cities are built. In fact, sinkholes are part of a broader problem of land subsidence, which simply refers to the ground sinking. Over time, subsidence damages buildings, roads, and transportation infrastructure. While individual sinkholes affect a single location, subsidence can occur over a large area—such as an entire city—over varying periods of time. For example, it is well known that Venice is sinking. This is driven primarily by the withdrawal of water, as well as compaction from centuries of building. Venice, however, is not alone in this regard. Mexico City is sinking due to water extraction from its underground aquifer, and Jakarta is sinking even faster. In addition to draining underground water sources, subsidence coupled with increasing sea levels could put Jakarta underwater in the not too distant future.

Sinkholes are not just reflections of geologic conditions; rather, they are a symptom of broader anthropogenic problems. For example, sinkholes can be the product of mismanagement of urban supplies or structural issues such as broken sewer pipes or water mains. In such circumstances, a sinkhole reveals a failure of policy or a political authority's stewardship of urban nature. Such a failure could be the by-product of inadequate support for infrastructure maintenance, ineffective land-use policy, or overly aggressive growth initiatives. Of course, anthropogenic climate change—tied to environmental stewardship— magnifies these failures as well.

An example of such mismanagement would be the unwillingness of policymakers to address environmental problems that lead to sinkholes. In 2016, the Mosaic Company, which mines and produces fertilizer, created sinkholes in

Massachusetts and Florida from its activities. Among the areas impacted by the company was a site outside Tampa that leaked millions of gallons of wastewater into another sinkhole before it was discovered in 2016. Despite spending millions of dollars to rectify the problem, the company's stocks are primarily tied to the price of their primary products: potash and phosphate. Thus, Mosaic suffered little direct financial harm. Without regulation to deal with real hazards, such catastrophes have little direct impact on corporations that create disaster risk.

The above case illustrates the way in which financialization can offset risk, as well as how the stock market is disassociated from social and environmental realities. Yet sinkholes are actually appearing, and cities are sinking. The sinking of cities, whether over long periods of time or through the catastrophic collapse of sinkholes, reveals a great deal about the nature of cities. Cities tend to be located near resources—especially water. Cities draw on local water sources for people, industry, and nearby agriculture and, ultimately, to grow. In addition to sinkholes, urban proximity to water supplies also means the potential for floods. Growth can also contaminate water supplies, leading to health hazards.

A desire for expansion-oriented growth coupled with the need for water creates the threat of sinking. Rapid urban development can weaken existing infrastructure, which can increase the likelihood of sinkholes. More people and businesses consuming water resources and more buildings and infrastructure being constructed in risky places have the potential to make cities sink. Population growth and economic activity also mean more fossil fuel consumption, which drives global climate change, resulting in rising sea levels and extreme storms that create additional hazards. Ultimately, sinkholes are at a nexus of human settlement patterns, poor urban planning, and resource consumption.

Ashley Dawson argues that urbanization projects are "capital sinks," which can in turn lead to the figurative and literal sinking of people and cities, whereby communities are left underwater.[1] Expansion-oriented growth, in other words, is central to understanding sinkholes and sinking cities. While cities have sunk in the past, contemporary sinkhole disasters are driven by the way capitalism encourages growth. Water and other resources are extracted to build cities. However, extraction for the purposes of expansion-oriented growth rapidly changes the landscape and creates new risks.

Studying Sinking Cities

Despite the social dimension of sinkholes and subsidence, scholars in most social science fields and urban studies have not examined them in great depth. Compared to other geohazards such as earthquakes and landslides, social scientists have generally ignored sinkholes. This is undoubtedly influenced by how disasters exist in our collective memory. Experience or awareness of events shapes topics of interest to social scientists. Sinkholes, much less slowly sinking cities, do not create the same fear and shock as do geohazards such as earthquakes or landslides. Typically, sinkholes are not "active" disasters such as floods, megastorms, and wildfires (which can rage on for days or weeks, allowing for endless news coverage). The news media has more difficulty generating what Timothy Recuber calls "empathetic hedonism" for sinkholes.[2] Damage caused by sinkholes can often be captured in a single static image; coverage is one and done. Moreover, the problem of sinking cities, like climate change, occurs over long periods of time and does not generate immediate emotional reactions.

In turn, research on sinkholes has primarily been conducted through the lenses of engineering, geology, or physical geography. This includes research on anthropogenic sinkholes caused by climate change, poor planning, population growth, and problematic resource management. The goal of these studies is to understand sinkhole formation for the purposes of prevention, early detection, or remediation.[3] Outside of those fields, limited work has been conducted on land use and urban planning. As urban disasters that cost a great deal of money to address, it is not surprising that sinkholes have also attracted some scholarship in business-related fields. For instance, there are analyses by the insurance industry. This work has mostly adopted an economic approach to analyzing sinkhole risk, often considering real estate development and advocating mitigation plans.[4]

In terms of social science, sinkholes are interesting because they lie at the intersection of human behavior, urban environments, and environmental concerns. The importance of sinkholes and subsidence is visible in the work undertaken by local and national hydrological programs, geological survey organizations and entities such as the United Nations. For instance, the United Nations Educational, Scientific, and Cultural Organization (UNESCO) has been concerned with subsidence since the 1960s because of this phenomenon's impact on cities, quality of life, and World Heritage sites. In fact, UNESCO's Working Group on Land Subsidence is one of its oldest working groups.

As mentioned previously, disasters are both a cause and consequence of city-building, including the social, political, and economic forces that shape the built environment. Riley Dunlap and William Catton note the importance of the built and natural environment in social life. The built and natural environment undoubtedly influences human action.[5] An emphasis on the built environment helps us to view sinkholes not as accidents or purely natural disasters; instead, they are products of human activities that involve interaction between the built and natural environment.

Both figuratively and literally, the foundations of many cities are at risk. By foundations, I mean the way in which we think about the structure of a city and how that structure is connected to the ground beneath the city. When it comes to buildings, properly laid foundations distribute the weight of a structure equally and increase its stability. We can similarly say that properly laid social foundations allow for equity and social stability. Problematic land use, rapid development, and mismanaged infrastructure all create risks for modern cities' physical foundation. This risk also has a profound impact on people living in these spaces.

Earth, Water, and Sewage

The extraction of raw materials, the consumption thereof, and the generation of waste are part of a city's metabolism. When I refer to urban metabolism, I am referring to the way cities, like living creatures, transform raw materials into usable energy and waste. This process affects not only how cities "live" within a given environment but also how they are built.[6] In the case of urban sinkholes, they are the by-product of humans believing waste materials vanish when they enter pipes or are washed away after a storm. Water (or sewage) comes from someplace, and it all goes somewhere. Growing cities need to deal with increasing amounts of trash and sewage. In earlier time periods, this led to sinkholes being used as dump sites. For example, Renaissance architect Leon Battista Alberti noted that sinkholes are perfect for sewage disposal because they devour and swallow everything and in turn reduce smells.[7] However, we now know that things do not just disappear.

A city's nature or metabolism is undoubtedly linked to urban nature or the rhythms of that city. In a living a city, soil and water, along with the built environment, interact, creating urban nature. There are several ways in which urban nature can shape sinkhole risk. The act of building cities and the creation of a built environment inevitably disrupt the earth and generate some form of risk. Coupled with the mismanagement of water supplies, such activities dramatically increase existing risk. An example is sewage and storm runoff that feeds back into urban nature—including soil, and bodies of water.

In the United States, areas that regularly encounter sinkhole problems include Alabama, Florida, Kentucky, Missouri, Pennsylvania, Tennessee, and Texas. Not only is karst common in these areas but humans have also settled near underground water supplies and other natural resources. As populations increase and people interact with nature, sinkhole risk increases. As such, cities find themselves vulnerable to this disaster. Given Florida's steady growth in recent decades, it is now the country's third-most populous state. Water management issues linked to consumption and hurricane-related flooding have made sinkhole prevalence more visible in Florida than in other states.

Similarly, rapidly growing cities in China see thousands of sinkholes occur each year. China has witnessed dramatic urbanization over the past few decades. In 1990, 26 percent of its population lived in cities; in 2010, the share of its population living in urban areas reached 45 percent. In addition, twenty-five of the world's 100 largest cities are in China. Given the urban development in China, sinkholes affect a significant portion of the country's urban population. According to Lei, Gao, and Jiang, sinkholes are a problem in thirty of China's major cities. Of the major sinkhole catastrophes in China, the authors note 75 percent are caused by human activities such as shoddy construction, foundation engineering, and water management.[8]

More broadly, urban growth in Asia has led to problems regarding sinkholes and subsidence. According to World Bank data, the percentage of the population living in East Asia doubled between 1960 and 2017, with 1.3 billion people living in cities. In addition to Chinese cities, Bangkok, Ho Chi Minh City, Kuala Lumpur, Manila, and other rapidly growing cities have experienced a wide range of engineering problems due to the presence of karst, groundwater exploitation, and rising sea levels. In Fukuoka, Japan, a massive fifteen-meter-deep sinkhole that spanned five lanes of traffic disrupted utilities and forced evacuations. A week after repairs were hastily completed, a new sinkhole opened up on the same site.[9] Despite how widespread these problems are, rapid urban growth is at the core of the problem. This expansion results in an ever-changing physical environment that is also becoming riskier.[10]

The South African Council for Geoscience notes the prevalence of sinkhole and subsidence in the vicinity of the rapidly growing cities of Johannesburg and Pretoria in Gauteng province.[11] For instance, the 1962 West Driefontein Mine sinkhole outside of Johannesburg raised sinkhole awareness in the country. Yet, as a result of rapid urbanization, an increasing number of people are now directly impacted by sinkholes. The growing severity of this problem has, in turn, magnified the inaction of politicians. The areas of Khutsong and Merafong in the West Rand district outside of Johannesburg have persistent problems with sinkholes. Rapid peri-urban growth has resulted (at least in the eyes of some locals) in politicians deprioritizing fixing existing problems in favor of new development.[12]

Guatemala City has also been the site of significant sinkholes. Like the cities described previously, Guatemala City has also grown rapidly in recent years and is currently the largest city in Central America. However, investigators have cited lax enforcement of building regulation and poor planning as exacerbating the city's history of sinkhole problems. Long-standing problems with a sewer line, in addition to the geology of the region, have made it particularly expensive to address the issue of sinkholes. In 2007, a 100-meter-deep sinkhole opened following a sewer failure and significant rain. Residents had reported noises prior to the collapse, but their reports were not taken seriously. Another sinkhole, ninety meters in depth, occurred in 2010 following Tropical Storm Agatha.[13] In such cases, policymakers were not unaware of the existing sinkhole risk because of the city's history of sinkholes.

Indeed, sinkholes, like other disasters (such as wildfires), occur naturally. However, like wildfire, human activity—especially city-building—can increase both the likelihood and the scope of the resulting disaster. City-building is a complicated engineering feat that requires builders to consider both the human and physical environment. The laying of foundations and building construction changes the physical environment in which the structure is situated. The site may have been relatively "natural" in that it was not used by people previously. However, in some cases, the ground a building is to be situated on may have been the site of a previous structure or was used by humans for some other purpose, including landfills. Either way, the construction of a building involves digging into the ground, running pipes, driving piles, laying foundations, and filling holes with compacted soil or other materials that will support the building's weight.

In 1922, British geologist Robert Lionel Sherlock referred to humans as a "geologic agent" that "disturbs the courses of rivers; fills lakes and makes new ones; checks or promotes sea-erosion; and modifies climates."[14] Sherlock noted that cities are places where humans have historically collected stone such as clay, shale, and sandstone to lay foundations and prevent sinkholes or subsidence. We often forget that all city-building, from the foundations to the structure itself, involves the movement of earth. This includes the use of fill material—consisting of rubble and soil—transported from other construction sites to create the ground beneath and around a structure. We establish quarries to mine clay that is molded into bricks and hardened in furnaces to build walls. Steel comes from iron that is mined, smelted, and mixed with other materials (such as coal) to create the skeleton of skyscrapers.

In turn, Ian Douglas speaks of cities as "dynamic earth."[15] The physical environment is ever-changing. However, some of these changes generate new hazards or exacerbate existing ones. To put it another way, the very act of building cities disrupts the earth in ways that creates risk. This poses problems for builders and engineers. Buildings, particularly large ones situated on vulnerable sites, require deeper piles for their foundations to be stable. This means that long, vertical post-like columns need to be driven into the ground to reach higher-density soil or rock for stability. In earlier time periods, wood piles were used for this purpose, but steel and concrete are used for today's megastructures. However, sometimes mistakes happen. For example, San Francisco's fifty-eight-story Millennium Tower, built in the mid-2000s, is currently tilting and sinking because of engineering flaws.

Another significant way that cities reshape their surroundings is to create infrastructure that moves water and waste in and out of a city. The history of the term *sinkhole* is itself fascinating in this context. Although similar holes in the ground can also be called cenotes, dolines, or swallets, the word *sinkhole* also has several social meanings. It commonly refers to a void or place into which things disappear. It has historically been used to refer to openings—whether in a home, a street gutter, or the ground—that could be used for disposing of waste or as sewage pits. The idea of waste is also present in the political use of the word to refer to wasting money. However, we know that things do not simply disappear—whether down a hole or in government programs.

While solid waste and sewage have been disposed of in water bodies, sewage is typically handled separately. For instance, ancient Greek and Roman cities had drainage systems and sewers as part of their infrastructure. In Greece, there were underground systems that diverted sewage and rainwater away from sources of freshwater.[16] Rome had the Cloaca Maxima, an extensive covered sewage system that diverted waste away from the city into the Tiber River. Solid waste, such as garbage, which is discussed more in chapter 3, was handled differently.

Similarly, modern sewage and municipal wastewater are collected through a system of pipes and brought to a treatment or disposal site. Carl Zimring points out that sanitary revolutions during the Industrial Revolution led to new means of dealing with waste—including new categorizations of materials, such as "wastewater" and "sewage."[17] While wastewater refers to all water that has been used and needs to be disposed of, sewage specifically includes human waste. To deal with this material, cesspools (or pits) and sewer systems are physically built into the material foundations of a city. As cities developed increasingly complex sewage systems in the nineteenth and twentieth centuries, sinkholes still served as discharge sites in many cases.[18]

Cities are relatively immobile; as such, they must rely on existing geography to cope with increased sewage and waste production. The use of sinkholes as disposal sites was a common practice as people began to move away from direct discharge of waste into rivers and other fresh water bodies. Historically, when one discharge source became unavailable, cities often turned to other nearby sites to deal with waste. For example, the town of Glen Falls in New York State had to rely on a nearby sinkhole on private property when they could no longer discharge their sewage into the Hudson River in the 1890s.[19] Similarly, the growth of State College, Pennsylvania, and development of what would become Pennsylvania State University in the early twentieth century involved the use of naturally formed sinkholes nearby to deal with sewage. In time, some sinkholes were filled and built over. This includes the current site of Memorial Field, which hosts the local school district's sports events and continues to be plagued with sinkhole-related damage.[20]

That said, the use of sinkholes as discharge sites was often contentious. The Mill Creek in West Philadelphia had a long history of sewage dumping and building that has since created subsidence and sinkhole risk. The creek's environmental history provides a prime example of how industry, urban development, and water management created perpetual sinkhole risk. In the 1890s, local officials in Philadelphia engaged in an active debate as to whether the mouth of the Mill Creek, which was already a site of heavy industry during the nineteenth century, was a sinkhole or part of a navigable stream. This debate was driven by the desire by railroad interests seeking to fill in the area to avoid bridging the stream.[21] Eventually, the creek was transformed into a sewer pipe and buried; over time, buildings were constructed over and around the area. This decision resulted in sinking homes and persistent sewer collapses and sinkholes throughout the twentieth century.[22]

Indeed, cities faced blowback for such so-called sewage solutions. It did not take long for city builders and public health officials to discover contaminated groundwater. After sewage overflows, there were legal battles and concerns regarding public health. Again, in Glen Falls, the reliance on a sinkhole for sewage led to landowners' lawsuits in the early 1900s. In the early twentieth century, public health officials and planners in the United States began to discourage the practice of using sinkholes for dumping sewage and other waste. Health bulletins explicitly warned that discharging sewage into sinkholes would pollute neighboring wells and suggested filling such sites.[23] A 1907 government report on typhoid identified the use of sinkholes along the Potomac as cesspits as a cause for disease outbreaks.[24] Even in the 1940s, there were typhoid breakouts in Minnesota because of towns discharging partially treated sewage into a sinkhole.[25] However, because of the American government's decentralized nature, formal regulation of sinkhole dumping is mostly a late-twentieth-century phenomenon. As a result, it is not uncommon to find cases of severe contamination of water supplies from sinkhole dumping well into the twentieth century.

Currently, some sinkholes are still used for water disposal (e.g., stormwater). Under federal law, these sinkholes are classified as Underground Injection Control (UIC) Class V Wells or "improved sinkholes." Like other sinkholes, they are natural-forming karst depressions. However, they have been modified to include some form of filtration, and are regulated. The primary goal is to protect drinking water. Despite these improvements, cities are often unable to deal with nonpoint runoff—or pollution from many sources—into sinkholes. As such, groundwater supplies may become contaminated.

Managing dirty water and runoff is a serious problem for modern cities. Sewer systems handle waste from within buildings by channeling it underground, whereas storm runoff comes from outside buildings. Historically, runoff was handled by combined sewers that handle both forms of wastewater. However, combining these systems led to overflowing sewers following a storm. In turn, more recent urban planning has called for separate sewer systems, also called municipal separate storm sewer systems (MS4) in the United States. These systems transport untreated polluted runoff into a body of water that is separate from a city's sewage system. In the United States, these systems are regulated with permits issued by the National Pollutant Discharge Elimination System. The issuing of such a permit requires that a city minimize harm caused by pollution. Such means can include the building of bioswales, which are drainage ditches filled with material that can filter pollutants in stormwater. While such strategies are increasingly common worldwide, cities still have difficulty building sewers and treating runoff.[26]

While modern sewer systems are tasked with collecting, treating, and disposing of wastewater, there are still problems in effectively keeping sewage out of sinkholes.[27] Even effective sewer systems that do not lead to sinkholes can break or crack over time. In addition, many cities around the world have aging infrastructure. Historically, pipes have been made of concrete, clay, wood, fiber mixed with tar (called Orangeburg pipe in the United States), cast iron, lead, and plastic. Sinkholes still occur because of the failure of sewage systems or water mains and after heavy rain.[28] Hurricanes and storms magnified by climate change generate persistent risk despite improvements in infrastructure.[29]

In addition to the increased sinkhole risk created by structures beneath cities, above ground structures can also create risk. This is because cities consist of many impervious surfaces that send water in many different directions. Buildings interact with rainfall, snowfall, and sunshine, which affect the social and natural environment. Anthropogenic climate change has created increasingly erratic weather conditions, such as heavy rainstorms and flooding. Impervious surfaces such as buildings and pavements, alongside poorly designed drainage systems, can result in subsidence or a sinkhole when storm runoff dissolves karst.

In other words, human intervention in the landscape and environment generates risk. Impervious surfaces can increase the risk of subsidence and sinkholes. Also, climate change can exacerbate subsidence. Ironically, water-related problems also include the absence of water. The Swiss Reinsurance Company (commonly called SwissRe) issued a report noting that heatwaves and droughts have substantially increased claims concerning subsidence-related damage throughout Europe.

Warmer and drier weather has made clay-rich soils more vulnerable to subsidence. In most cases, the result is minor cracks in walls, broken doors, and windows. However, it can also mean more substantial foundation damage, especially for homes built on clay along major rivers. The report notes that the areas along the River Thames are particularly susceptible following a major heatwave. In 2018, London saw a 20 percent increase in subsidence-related damage from the previous year. Furthermore, Europe has faced regular record-breaking heatwaves in recent years. Countries, such as France, have experienced a 50 percent increase in heatwaves over the past twenty years.[30] As such, it is generally argued that land-use approaches be adopted that bear in mind sinkhole risk and preventative measures.[31]

The Politics of Sinking

As cities grow, public agencies must invest heavily in developing infrastructure and manage the increased risk of that infrastructure breaking down. Investment or funding includes covering damage from sinkholes. However, sinkholes are often associated with expensive government projects—even if they focus on fixing sinkholes themselves. As noted previously, the term *sinkhole* is also used to refer to failed (primarily government) projects that waste taxpayer dollars.[32] For example, a 1914 Texas report on street paving inquired if it was "any wonder that, under these conditions, the streets of the American city have been a *sinkhole* in which, constantly, dollar after dollar of taxpayers' money has found its resting place?"[33] In 1924, *Public Service Magazine* gave the example of a "situation in Chicago, in the case of the sanitary district payrolls. This sanitary district, it will be recalled, comprises the great municipally-owned and operated electric system, which for years has been a tremendous sinkhole for the taxpayers' money." In the 1940s, commentators expressed the concern that New York would find itself falling "into the bottomless sinkhole of 'pressured' budgeting."[34] To this day, the term "sinkhole" still has this negative meaning. Yet it is often up to government agencies to repair the damage caused by actual physical sinkholes and other disasters. Ironically, government agencies are both blamed for sinkholes and failures in fixing them. Correspondingly, this chapter concludes with a discussion of "sinkholes" for public policy in an era of neoliberalism and disaster capitalism.

Of course, not everyone believes infrastructure projects are sinkholes. Daniel Webster Hoan, the socialist mayor of Milwaukee, Wisconsin, defended government projects in the 1930s by noting that "some believe that government is a sort of '*sinkhole*,' that all taxes paid in somehow disappear. It is not recognized that practically all of the money paid in taxes goes into payrolls and for materials, and hence is quickly placed in circulation."[35] In reality, effective urban governance is essential to building resilient and less risky cities. In the late nineteenth and early twentieth centuries, American cities saw the construction of complex drainage, filtration, and treatment systems to deal with runoff and wastewater. Since then, there has been ongoing work to develop new infrastructure types that can safely manage stormwater and sewage. In the 1990s, Seattle integrated municipal management of its freshwater supply, stormwater, sewage, and solid waste management to better address the city's interconnected environmental problems.[36] While the deployment of such strategies is rarely straightforward, it is clear that government projects are not necessarily sinkholes or a waste of resources. The problem is that effective and resilient planning is invisible when it functions correctly; in turn, this makes it easy for neoliberal interests to claim that taxpayer money has been squandered.

Sinkholes caused by sewer or other plumbing breakdowns can affect municipal water supplies, forcing residents to boil water until repairs are completed. Sinkholes also create obstructions in the urban built environment that may make life more difficult, in part because water and sewage infrastructure often run under or alongside roadways. For example, in early-twentieth-century New York, sinkholes made various parts of Broadway unpleasant and painful to navigate.[37] It took about two years to repair Seattle's infamous 1957 Ravenna sinkhole. More recently, in Fraser, Michigan, a sinkhole created by a sewer line collapse took nearly a year to repair. Similarly, it took years before political leaders in Harrisburg, Pennsylvania, allocated funds to fix long-standing sinkhole and subsidence problems.[38]

Despite this disruption of everyday urban life, politicians concerned with city budgets often downplay the significance of these problems. In 2017, Canadian politicians in Ottawa made a concerted effort to use terms such as "road degradation" and "pothole" before calling the situation a sinkhole.[39] The implication is clear, as a pothole in the road is smaller and less expensive than a sinkhole. Fixing a sinkhole requires an assessment of the problem and potentially expensive repairs. The hole needs to be filled, the pipes need to be repaired, and the street repaved. Given the layers of bureaucracy and the effort involved, it is possible that a repair will either be rushed or that it will take years for politicians to secure the funds needed to repair sinkholes.

In the contemporary world, neoliberal policymakers' de facto response has been to mitigate the short-term harm caused by calamities. While mitigation strategies can include enacting regulations or implementing engineering solutions such as compacting existing soil, excavating and refilling sites, and reinforcing infrastructure, these strategies do not eliminate the risk of sinkholes. It is not possible to merely retrofit homes to be sinkhole resistant or build large retaining walls to deflect karst-dissolving water. The only way to deal with sinkholes is either by limiting construction or creating regulations that protect water sources. In addition to the cost of sinkhole-related damage to the built environment, there still is also the problem of dumping waste into sinkholes, which can contaminate groundwater.

Laws regulating discharge into sinkholes were eventually enacted in the late twentieth century. While the U.S. federal government introduced the Clean Water Act in 1972, the law did little to regulate nonpoint pollution or address waste discharge into sinkholes. While states could regulate discharge into sinkholes using the National Pollutant Discharge Elimination System, laws vary significantly by state and municipality. For example, the regulation of sewage in karst terrains in Pennsylvania did not begin until the mid-1970s.[40] Alabama passed a cave protection law in 1988 that restricts discharge into sinkholes. This fragmented approach to clean water, sinkholes, and subsidence is, of course, problematic given the interconnected nature of hydrological systems.

While proper planning to divert runoff and to maintain infrastructure can reduce the risk of sinkholes, urban growth and climate change are barriers to building less risky cities. As cities grow, driven by capitalistic impulses, risk increases. It is like trying to plug a dam that is cracking. Again, the plights of cities are tied to their geographical locations. Cities are generally built in relation to water resources and other amenities. However, such a locational advantage creates precarity.

Capital Sinks and Florida Sinkholes

In today's neoliberal city, we see the above-mentioned predatory nature of capitalism at work when natural disasters occur. Florida has received considerable attention because of the urban growth that it has witnessed, its renowned tourist destinations, and the volume of the sinkholes that have appeared in the state. Florida represents a convergence of climate change–magnified hurricanes, geology, and urban development.[41] Despite this, real estate developers have profited greatly while increasing sinkhole risk by building in Florida's vulnerable areas.

In addition to apparent sinkholes in residential communities, there are other issues. These include natural amenities such as lakes going dry as aquifers are drained by human use, which increases sinkhole risk. Another problem is that state laws only require the seller of a new home to disclose to a buyer whether a sinkhole claim has been made. Also, given the speed of development and sales, homeowners are the ones who ultimately deal with sinkholes, not builders. While large developments do include various landscape elements, such as bioswales, basins, or rain gardens, to help manage storm runoff, such elements cannot deal with the environmental changes from rapid urban development and hurricanes. As such, regulation, as well as restriction of ongoing development, is essential.

Statewide, there was a jump in total indemnity payments—or losses paid directly to insured parties—for sinkhole-related claims from $163 million in 2006 to $220 million in 2009.[42] Between 2000 and 2010, the Tampa Bay Metropolitan Statistical Area has grown by 26.5 percent. Counties outside of Tampa, namely Pasco, Hernando, Pinellas, and Hillsborough, comprise what is called "Sinkhole Alley." This area has seen steady population growth; at the same time, the number of sinkhole claims has dramatically increased. Also, counties around Orlando, such as Orange, Marion, and Lake, have seen a population increase of over half a million people over the past twenty years. Here, we have a feedback loop in which population increases lead to increased risk. This increases the number of sinkhole-related claims. The result is a dramatic increase in financial risk for insurers that assist in home repair or fixing infrastructure.[43]

While a great deal of research has been conducted on sinkhole risk in Florida, predicting a sinkhole does little to stop the human interactions that increase sinkhole risk. Typically, a great deal of capital will have been invested or sunk into the built environment. For instance, once a community has already been built, it is immobile. Houses are immobile and provide habitable spaces for people. As such, the impact of constructing homes and communities cannot simply be undone as buildings cannot be moved. Therefore, both buildings and people are exposed to persistent risk. Moreover, it appears that sinkholes swallowing whole homes does little to slow development. The existence of disaster allows new forms of revenue to be generated, as the financial sector has found ways to commodify disaster through insurance and offset economic risk with public entities' assistance.

Since 1954, the Florida Geological Survey has maintained a Subsidence Incidence Database to track sinkholes. However, there is no government agency tasked explicitly with sinkhole inspections in Florida. Before the 2000s, sinkhole legislation in Florida focused on water quality, not structural damage. It was not until the Water Resources Act of 1973 and the Water Quality Act of 1983 that statutes restricted discharge into sinkholes.[44] In 1982, the Florida Sinkhole Research Institute was created to conduct research and collect data on sinkholes. However, the institute was defunded by the state legislature in the 1990s. In recent years, there has been growing concern regarding sinkhole-related structural damage. However, Florida worked to expand and enhance insurance coverage rather than regulate development.

The adoption of this approach was driven by rising claims for sinkhole damage, as well as a high number of claims being denied. In 1992 and 2005, the Florida Legislature ordered analyses of sinkhole insurance. In 2002, the Florida Citizens Property Insurance Corporation was created by the Florida Legislature as a not-for-profit insurer for those who could not obtain homeowner's insurance. While the company was primarily set up to provide hurricane assistance, it also provided coverage for sinkholes. Over the years, Citizens Property Insurance has continued to expand coverage throughout the state.[45] This expansion is, in part, linked to Florida statute 627.706, which requires insurers to cover catastrophic ground cover collapse. At present, the only other state that requires insurers to provide coverage for sinkholes is Tennessee. These laws were driven by insurers avoiding sinkholes policies in counties within "Sinkhole Alley." At the same time, Florida law is very specific regarding sinkhole-related damage and whether such policies meet the legal criteria for ground cover collapse. Florida law limits mandatory coverage to catastrophic damage only, not claims for cracks or other damage that is deemed "minor."

While coverage has increased, the situation is similar to the National Flood Insurance Program, a last-resort insurer. Under the Flood Disaster Protection Act of 1973, flood insurance is mandatory for all federally backed mortgages.

There was further expansion of coverage under the National Flood Insurance Reform Act of 1994. The program was intended to make homeowners and communities more educated in terms of coping with flood risk. However, it ultimately led to destroyed homes in risky environments being rebuilt with taxpayer-supported insurance money. Simultaneously, national flood insurance coverage has increased four and a half times in the twenty years since the law was passed.[46]

Expanded coverage, without addressing the underlying causes of disaster, has meant the number of claims filed continues to increase. By 2010, Citizens Property Insurance was receiving over 200 sinkhole claims a month. The cost to homeowners through related increases in premiums at the time received widespread attention both within the state as well as nationally.[47] In response, Florida's legislature again modified state law to expand coverage, but interestingly, it also limited the burden of insurers. For instance, Florida Senate Bill (SB) 408 was passed in 2010 to redefine different types of damage and gave insurers the ability to limit coverage—for example, to only one's home, not the driveway or decks thereof. It also allowed insurers to create limits for partial losses and placed a two-year limitation on claims. These changes essentially gave Citizens Property Insurance more power to deny claims—which, in some cases, led to costly legal battles.[48] Later, in 2016, SB 1274 allowed private insurers to offer optional coverage for both the structure (i.e., a house) and personal property for higher premiums.

Of course, the primary disaster concern in Florida is hurricanes. Throughout the fall of 2017, Florida was recovering from the aftermath of Hurricane Irma. However, in addition to homeowners filing claims from hurricanes, there was also increased sinkhole activity driven by the deluge.[49] There has been some limited discussion of a state sinkhole fund, not unlike the existing hurricane fund, to help insurers in the same way that hurricane victims are assisted by a fund that also spreads risk to the marketplace.[50] Again, the reliance on public disaster funds and various financial instruments is intended to redistribute financial risk.[51]

Nonetheless, homeowners' insurance premiums have gone up.[52] The simplest explanation for this development is that insurance companies need to develop ways to redistribute the financial burden of those losses. Being unable to rely on premiums alone, in 2012, Citizens Property Insurance began investing in catastrophe bonds (specifically, the Everglades bond). A claim made by insurers and public agencies and politicians is that catastrophe bonds and other financial products reduce the potential harm that taxpayers might suffer in the case of a disaster. The idea was that the state would not have to rely solely on the Florida Hurricane Catastrophe Fund—a publicly managed trust fund designed to support insurers in the event of catastrophic losses.[53] This is an example of how neoliberalism does not mean the absence of the state. Instead, laws are created that support neoliberal policy, thus expanding the market while minimizing private companies' risk.[54] We must remember that the Hurricane Catastrophe Fund supports insurers more than the insured. Despite having paid out many claims in recent years, Citizens Property Insurance reported that it was in good financial shape, in part thanks to the catastrophe fund, as well as through reinsurance coverage and other means of mitigating risk.[55]

Engineering Solutions the Hard Way

It is perhaps not surprising that there has been growth in the sinkhole and subsidence "remediation" industry both in Florida and elsewhere. There also have been developments in NASA's interferometric synthetic aperture radar (InSAR) to detect subsidence and help predict sinkholes. The use of geographic information systems (GIS) and digital elevation mapping can also allow agencies to monitor subsidence.[56] However, none of these tools are preventative. Although they *could* discourage development in vulnerable areas, that is not what we have seen; instead, expansion-oriented growth continues. In the absence of aggressive regulations intended to reduce sinkhole and subsidence risk, the result is various businesses working in the broad area of sinkhole inspection and remediation. In Florida, the concern regarding sinkholes has led to engineering and surveying companies developing tests that use ground-penetrating radar to assess sinkhole risk for both developers and homeowners.

However, experts have questioned the usefulness of such tools because of hurricanes and ongoing urban development creating ever-changing conditions.[57] While sinkhole detection has created new business opportunities, the pace of development limits the usefulness of detection. With ceaseless construction, the physical landscape will have been changed within a few months. The escalation of risk results in more economic actors transforming this environmental *bad* into a revenue stream. Construction engineering/surveying companies, homeowners, government officials, insurance companies, risk managers, and, of course, banks and the financial sector have devised ways to profit from disaster.

Given the involvement of the actors identified above, there is a tension between risk mitigation and profit. Indeed, most of these actors are actively attempting to engineer harm away. From the standpoint of engineering and remediation, fixing the problem represented by a sinkhole is a reasonably straightforward physical process. The process mainly involves identifying the cause, plugging and filling the sinkhole, and then covering the surface. Doing so may entail mending broken pipes or removing the water source that produced the sinkhole. After that, the hole will need to be filled. This can be done through compaction grouting, which refers to adding material to plug the hole or change the soil's chemical composition. Finally, damaged infrastructure or buildings will need to be repaired.

It could also be said that such solutions are further technical interventions into the natural and built environment. While a broken pipe is relatively straightforward to fix, addressing runoff and other water sources is more complicated. Excavations intended to deal with water and fill sinkholes can range from straightforward drilling to the use of hydrofracturing, which can create new fissures. Filling a sinkhole involves digging and creating new human-made earth the building sits on, to replace what existed before. These interventions reveal the dynamic nature of the ground on which cities are built and suggest that the foundations of our cities are vulnerable.

Solutions to problems are not neutral. How we approach problems reveals the underlying belief systems of those involved and society's value systems. Today, we continue building cities in the hope that we can design or engineer our way out of hazards such as sinkholes. However, this approach does not consider human or social drivers of risk. For the most part, these solutions could be described as "hard engineering" solutions as opposed to "soft" or social solutions. Historically, the idea of hard engineering applies to measures such as coastline management strategies—for example, the building of seawalls and levees to prevent rising sea levels from harming cities, preventing seawater from contaminating freshwater sources, and making cities more resilient to storm surges. Such large construction projects result in permanent physical features, such as a massive wall, being inserted into the environment.

Hard engineering is often seen as a solution to sinking cities. Before its cancellation, Jakarta's $40 billion Great Garuda Sea Wall project was expected to prevent the city from sinking further. In the meantime, the construction of smaller sea walls has continued in the hope of mitigating some of the city's subsidence. Other engineering solutions include the creation of new human-made infrastructure to replenish aquifers to slow the rate at which it is sinking. In the case of Shanghai, the government is pumping water underground to offset the city's sinking problem, while Mexico City has investigated developing pipelines to bring in water from a source far away from the city. These measures, not unlike the approaches adopted for dealing with smaller sinkholes, fundamentally rework water and earth in the hopes of slowing or preventing subsidence.

As with sinkholes in Florida, the problem with such solutions is that they cannot adapt to future climatic or population changes. They are, after all, hard and immobile interventions into environment. They are not necessarily flexible or adaptable should further changes occur. In fact, hard engineering solutions are often built with the assumption that nature—which includes rivers, plants, and wildlife—does not change or vary. On the contrary, ecosystems and human settlements move, grow, and shrink.[58] As sea levels continue to rise and people increase water consumption, the risk to human settlements will increase.

The debate over hard engineering is not unlike debates over other techno-fixes, such as geoengineering. Geoengineering typically involves the development of technologies for managing solar radiation or removing carbon dioxide to mitigate the effects of anthropogenic climate change. Advocates argue that we have already crossed the tipping point where human civilization can reverse the harm that we have caused to the environment by merely reducing emissions. This strategy has been championed by celebrity entrepreneurs, corporations, and neoliberal policymakers as perhaps our only hope of slowing or turning back the clock on the damage we have created. Their support is tied to the investments being visible; investors can participate in groundbreaking and ribbon-cutting events. Such approaches are often further interventions into the built and natural environment that create new disruptions by offsetting the disaster, as opposed to dealing with the causes thereof.

Therefore, Ryan Gunderson, Diana Stuart, and Brian Petersen argue that geoengineering does more to reproduce capital's obsession with accumulation than to work toward a stable climate system.[59] They have criticized the technological rationality used to legitimize geoengineering. Specifically, they note that decisionmakers have framed geoengineering as a viable "Plan B" to the "Plan A" of reducing greenhouse gas emissions. However, the removal of carbon or deflecting solar radiation is not the same as moving away from an economic and social system that relies on fossil fuels. Geoengineering "masks and reproduces capital's contradictory need to self-expand, on the one hand, and maintain a stable climate system, on the other."[60]

These interventions are deployed after sinking has already begun or, in the case of geoengineering, warming has already begun. Correspondingly, environmentalists and scientists have raised concerns that such programs may disincentivize real change. Rather than dramatically changing the way we think about urbanization or reducing our overall production of greenhouse gases, we will design away human-driven problems. This approach represents a form of anthropocentric hubris. This overconfidence in human ingenuity makes such solutions blind to human factors such as capitalism that exacerbate disaster risk.

This anthropocentric hubris also has potential blowback. A real concern regarding geoengineering is the unforeseen consequences of directly intervening in the atmosphere. The co-option of various strategies to cope with disaster may lead to unintended consequences for cities. Often, disaster mitigation and prevention strategies create new, unexpected harm.[61] For instance, waste-to-energy approaches, as a seemingly sustainable solution for the problem posed by trash that conveniently produces electricity to drive our consumption habits, fuels climate change. Fire-resistant architecture has the improbable goal of creating safe homes. However, the repercussion of its deployment in the wildland–urban interface is ongoing development in a flammable environment. The establishment of sewer systems to handle our wastewater has improved public health and allowed for ongoing urban growth. However, it has also increased the risk of sinkholes.

In conclusion, the science behind sinkholes is not new. It is quite well known that the mismanagement of water systems, problematic land use, aquifer compaction, and pumping are causing cities to sink. However, subsidence is affecting more extensive regions of the world as they urbanize. As UNESCO notes, subsidence "is a major problem that threatens viability and sustainable economic development for many millions of people throughout the world."[62]

While cities are relatively fixed in terms of space, disaster consequences are often not limited to a single place. The displacement of human beings, financial loss, and the breakdown of intercity connections indicate that environmental vulnerabilities are shared. Therefore, sinkholes are not just isolated incidents within neighborhoods; rather, they are by-products of growth and expansion—both of which are essential parts of capitalist ideology—that drive cities to externalize both costs and risks. We thus need to think about what really constitutes urban and economic resilience. While mitigation strategies such as using new technology to predict and warn residents of sinkholes offer *some* protection to homeowners who can afford them, we must be aware of how contemporary capitalism has structured the relationship between the built environment and nature. Thus, any analysis of future policies and solutions needs to tackle the risky foundations of cities, as it is only by understanding these relationships that can we better protect communities from disaster.

3

The Logistical Nightmare of Trash and Urban Nature

• •

> The mountains of things we throw away are much greater than the things we use. In this, if in no other way, we can see the wild and reckless exuberance of our production, and waste seems to be the index.
> —John Steinbeck, *Travels with Charley: In Search of America*

You might be wondering how garbage fits into a study of urban disaster. The handling of municipal solid waste is critical to the functioning of cities. Urban areas—because of their population density—are persistently at risk of an environmental crisis because of waste. In early 2018, the European Union threatened to sanction Italy for the waste crisis in Rome. Italy had already been sanctioned in 2010 by the European Court of Justice for trash problems. At the time, it argued that Italy had not sufficiently implemented EU laws on waste to the point of creating health hazards.[1] For decades, Puerto Rico has faced severe problems with garbage collection and landfill capacity. Following Hurricane Maria in 2017, waste become a full-blown disaster because of inaction by the U.S. government. Also, in 2017, attention was brought to the ongoing garbage problems in Lebanon that created domestic health hazards and pollution in the Mediterranean. These are just a few examples of the mishandling of garbage and how it leads to a crisis. Using garbage as our next case of urban risk gives us insight into the problem of urban growth and how growing cities offset the risk of such crises by turning waste—an environmental *bad*—into a potential *good*.

Before going further, it is essential to note that this chapter deals with municipal solid waste (MSW), garbage, refuse, trash, or rubbish. This is different from sewage or wastewater discussed previously. Indeed, there are times—in both the past and present—where both solid waste and sewage were disposed of in the same bodies of water. However, sewage is typically categorized and handled separately. Trash is typically handled above ground, and sewage is dealt with through underground pipes. The vulnerability of waste infrastructure has long been recognized by social scientists and urbanists. In the 1930s, Stuart Chase suggested that modern urban infrastructure is highly vulnerable to disruptions, or in his words, "a central nervous system lying almost naked for the cutting." While Chase lists transit, telephones, and other examples of infrastructure breakdown causing chaos, he notes that garbage collection was "a very tender nerve." This vulnerability, he argues, is exacerbated by private interests, which "are continually capitalizing these inventions and introducing them into the industrial structure, often with the formula kept secret."[2] Chase's commentary nearly a century ago reveals the way that the breakdown of garbage infrastructure is not only a potential disaster but how its evolution is connected to the development of capitalism.

Indeed, waste management is not merely a question of infrastructure. Instead, an examination of trash reveals the way corporations benefit from risk and disaster capitalism. For instance, Waste Management, Inc. (WM) is North America's largest waste management firm, far outpacing the second largest company, Republic Services. WM continues to grow. It has acquired Anderson Rubbish Disposal and Moorpark Rubbish Disposal in California, Illini Disposal in Illinois, and Pioneer Industries in Wisconsin, to name a few companies it now owns. In 2009, WM purchased 40 percent stake in the Shanghai Environment Group (SEG) that provides waste transfer, handling, and landfill services. Since the market crash of 2007 and 2008, WM's stock has fared well on the New York Stock Exchange, thanks to increasing revenue and growth, and all because we keep generating trash.[3]

In other words, waste management is highly profitable. However, it is not just the management of waste. This is about the transformation of so-called waste into something of value—a commodity. Michael Thompson proposed that the category of "rubbish" or "waste" is not fixed. He argues that seemingly worthless items that are thrown away can become new commodities called antiques or scrap. This means the value we place on so-called rubbish can change. Garbage can become a resource.[4] There is no doubt that contemporary capitalism has commodified various forms of trash. For instance, garbage in incinerators becomes fuel, which provides more energy for people to consume. The recycling of plastic results in a market for postconsumer resin and the production of new goods for people to buy.[5] This is seen as better than dumping trash in landfills, but the emphasis on fueling consumption suggests that sustainability is not primarily the goal. While reducing harm to the environment and human health is appreciated, the potential profit created from reclassifying rubbish serves growth-oriented capitalist ideology that undermines sustainability.

This chapter approaches garbage in two ways. The first is to discuss the role of urban infrastructure development and its impact on another city's development. Cities, after all, influence one another. They not only export waste to other cities, but they export waste management services and infrastructure development. Indeed, the export of hazardous wastes from wealthy to developing nations is well studied.[6] Yet the flow of waste management and logistical services between newly industrialized countries, lesser developed and peripheral countries, or South–South relationships are underexamined.[7] Here, I want to move the discussion away from just wealthy cities in the Global North exporting waste to the Global South and target urbanization, growth, and expansion. Not merely driven by population increases and sprawl, cities need to efficiently manage their consumption and waste production in order to grow—or else face crisis.[8] However, growth creates a contradiction, whereby increased consumption leads to more waste—a potential vulnerability. Put more simply, risk and precarity are linked to waste. Trash hurts the environment, has an impact on public health, and threatens the everyday functioning of cities.

Cities, as we saw in chapter 1, are stuck in place. As they grow, they produce more and more waste and have fewer and fewer places to put it. To deal with this problem, cities increasingly rely on other cities for economic expansion (e.g., markets) and to sell services such as metropolitan solid waste management. As such, this chapter focuses a great deal on logistics. Logistics, simply put, is the management of the flow of goods and services. Logistics enables higher productivity as demand for natural resources increases through efficient production, distribution, and rapid consumption. For this to occur at larger and larger scale, the removal and handling of waste requires more complex and integrated global waste management systems. This is not just happening in wealthy countries in the North. Cities in newly industrialized countries (NICs) are not just developing strategies to handle solid waste domestically, they also are exporting services overseas.[9]

In turn, the second issue this chapter addresses is the marketization of trash and its role in fueling urban growth. This means that the development of infrastructure to handle trash and the development of technology such as incineration or the industry term waste-to-energy (WTE) have effectively transformed *waste* into a *good*. It is a commodity. Trash becomes a fuel source in this system. However, it is not just municipal solid waste that becomes a commodity. The ash produced by incineration has also become a commodity, as it can be used in the production of cement—a building block of modern cities. The transformation of waste management into logistics, for the time being, helps to offset the risk of too much waste or running out of waste to fuel incinerators. While logistics allows for growth predicated on increased consumption, it also redistributes the risk associated with trash. It facilitates the movement of garbage and related risks from cities to other places such as landfills, incinerators, or other countries.

A Brief History of Metropolitan Solid Waste Modernization

From ancient to early modern times, trash was often thrown out into the street. Before modern regulation of waste, there was still a high level of informality when it came to garbage disposal—such as dumping in waterways (ocean, rivers, and wetlands), open pits, and backyard incineration.[10] Ancient Roman cities had sewers, such as the Cloaca Maxima, that were part of their infrastructure. However, solid waste collection was not. While sewage was washed away in a sewer, trash was an everyday threat to public health as it was thrown out into the street. Roman poet D. Iunius Iuvenalis, also known as Juvenal, joked that one would pray and hope that "local housewives drop nothing worse on your head than a pailful of slops" from their windows.[11] When the garbage thrown out on the street was disposed of, it was transported through aboveground routes to the outskirts of the city where it was buried in open pits or burned.[12]

As a result of industrialization and the growth of large urban centers, cities needed more efficient ways to deal with streets blocked by debris, trash, and hazardous materials. This filth could affect urban health and growth—limiting economic activities and lead to crises such as cholera outbreaks. As early as the eighteenth century, cities such as London were establishing mechanisms for the disposal of such material. In 1751, London mayor Corbyn Morris argued that the "cleaning of the city should be put under one uniform public management."[13] By the end of the nineteenth century, public agencies were becoming commonplace in large, industrialized cities. In addition, landfills and incineration became common ways to reduce trash volume and disease. However, it was not until the twentieth century that it was entirely accepted that waste management is a municipal responsibility or a public good that required government involvement.

It is important to note here that the rise of formal municipal waste collection was not instantaneous. In early modern England, toshers, mudlarks, and dustmen were collectors and recyclers of various materials that facilitated the removal of trash in exchange for profit. Independent "entrepreneurs" hauling waste to the urban periphery or collecting scrap have long played a role in municipal solid waste collection and recycling. Today, there are still informal collectors of recyclables in American cities. Elsewhere, itinerant waste collectors and scavengers in waste dumps play an important role in many developing countries—alongside formal waste management. They fill the gap between various forms of formal waste collection. *Chamberos* in Ecuador, *zabaleen* in Egypt, *pepenadores* and *chatarreros* in Mexico, and *hurdacı* and *eskici* in Turkey are just a few examples of collectors and recyclers around the world.[14]

Structurally speaking, industrialization and urbanization are important in the development of formal MSW management because relative "plentifulness" allowed for a consumption-oriented throwaway culture.[15] As critical environmental scholars have argued, "affluence" is a significant variable in measuring environmental impact. Citizens of wealthy countries generate significantly more waste than their counterparts in the developing world. Cities—especially those in the Global North—are densely populated sites of consumption that have, in many cases, limited space to grow. In turn, they have a higher level of urgency in dealing with waste than less densely population centers.[16] However, we are seeing this become important in newly industrialized countries. As such, MSW regulation and management is now an integral part of municipal public policy in cities around the globe.

MSW companies vary in their organizational structure and charter. In contemporary cities, they consist of private companies, large companies with the national government as a major shareholder, or municipal public benefit corporations. Whether private or public, the development of widespread MSW services is seen as good public policy. However, private models, in particular, reflect a city's fiscal policy. In many cases around the world, systems are financed through built-operate-transfer (BOT) or build-own-operate (BOO) models. This is a means of allowing a private company to build facilities that provide a public good such as MSW management. In exchange, the municipality may give the private company incentives or concessions, such as tax benefits or future profits generated from the facility. The city may also provide funding, or the private company may be required to finance construction and operations for the duration of the contract.

In addition to providing MSW services to single cities, some agencies and companies operate regionally, nationally, or even globally. Diane Sicotte notes that the Philadelphia metropolitan region became a magnet for garbage coming from New York with the help of firms such as Covanta.[17] This is in part the result of a changing urban hierarchy within the United States and around the world.[18] For instance, American Rust Belt cities increasingly prop up the development of cities such as New York and companies based there. Metropolitan

Recycling in New York City, for example, proudly notes on its website, "Metropolitan is poised to expand through acquisitions and development projects including our newest venture located in Detroit, Michigan."[19] However, it is not just American firms. The largest metropolitan solid waste company in Europe is the French firm Veolia (which was sued for its role in the Flint, Michigan, water crisis).[20] Major Western European firms such as Veolia are especially present in later EU ascension countries such as Croatia and Romania with BOO contracts, illustrating expansion into new urban markets.[21]

It should be noted that cities collaborate on infrastructure, and waste is no different. While national policy regarding intermunicipal cooperation varies, we should recognize that the convergence of the world's cities on the use of large firms to dispose of waste is not merely one of utility and policy transfer.[22] Indeed, cities transfer policies and technical strategies to one another. However, the replication of policy and practice is influenced by unequal social relations and global processes, as well as the emergence of hegemonic understandings of urban governance.[23] In the case of MSW and other green strategies, Bunnell and Das describe them as "seductive"—given their marketability.[24] With the rise of neoliberalism in the Global South, "green" policies are just as popular as they are in the North. This is because going "green" is especially seductive when paired with discourses linked to modernization and global competitiveness.[25]

Waste management becoming a municipal responsibility, or else one that is outsourced, did not mean that the problems associated with trash were solved. Rather, risks were deferred. Cities initially hauled garbage out to the outskirts of cities, but as cities grew, former landfills became the physical foundations of new buildings and parks. This, of course, has been tied to environmental inequality and racism. In fact, whole neighborhoods in New Orleans were built atop landfills throughout the twentieth century, with long-standing implications for the residents' health and safety, as well as the creation of sinkhole risks.[26]

As new transportation technologies were developed, the ability to move trash further and further away from the city became possible. Yet, it still had to go somewhere. As late as the 1980s, there were high-profile garbage crises. In 1986, there was the *Khian Sea* garbage barge incident, and in 1987, there was the *Mobro 4000* incident. New Jersey refused to allow the *Khian Sea*—loaded with ash from incinerators in Philadelphia—to dock. For over a year, the ship went from port to port, from the Caribbean to Africa and even Asia to find a place to dispose of its waste. The material was eventually dumped in the ocean. Like the *Khian Sea*, the *Mobro 4000* left Islip, New York, for North Carolina but was turned away. The barge bounced from port to port, going as far south as Belize before returning to New York.

As a response to such incidents, the Basel Convention was signed in 1989 to limit the shipping of hazardous waste. Despite the agreement, potentially dangerous waste is still shipped around the world—especially when it is reclassified as a commodity. Generally, the externalizing of waste production is something we observe among wealthy countries in the Global North. Derek Kellenberg suggests that when consumption creates a negative externality and trade costs are asymmetric, then one sees the shipment of waste North to South. Put another way, when increased consumption levels begin to have negative consequences, countries take advantage of uneven development to export waste.[27]

There is no doubt that urban growth benefits from the externalization of wastes and environmental hazards by taking advantage of inequality around the world. It should be noted that movement of waste is more complicated than just the reliance on developing countries to take American and European waste. NICs and their cities play an integral role in commodity chains and, ultimately, the production and export of wastes. It could be argued that this is an example of the ecological modernization of cities in NICs. As cities in the Global South "modernize," they mirror the exploitative processes found in the North.

According to ecological modernization theorists, such crises provoked significant changes in waste management.[28] Recycling, for example, became a potential solution to overfilled landfills and fears of hazardous waste from incinerators. As Pellow, Weinberg, and Schnaiberg have argued, it was a plan that urban policymakers found palatable and was a potential source of revenue.[29] The rise of recycling, as well as the general history of metropolitan solid waste management, is not unlike other stories of urban development. It is an example of solving problems through technology, growth, and other market-driven solutions. Yet, ecological modernization is typically a form of neoliberal adaptation. Cities rely on private companies to handle waste and these contracts are financially lucrative. However, since recycling does not directly deal with the problem of consumption, it is a coping mechanism for waste production rather a solution.

For instance, China's refusal to import trash and recycling has created tensions with other countries. In 2017, China formally informed the World Trade Organization (WTO) that it would no longer accept imports of foreign garbage. Previously, an estimated $5.6 billion a year was generated by the American export of scrap to China. In addition to losing revenue, Americans, Europeans, and other exporting countries have seen a backlog of waste. This move is indicative of China's ecological modernization, as it has made environmental issues a key component in its economic policies. As China's economy has grown, the prospect of importing environmental harm has become less desirable.

Such an approach to dealing with environmental problems represents a view similar to the Kuznets curve. This model of modernization suggests that when an economy develops, inequality or some other social problem grows before stabilizing and then declining. This model has been adapted to environmental problems, suggesting that economic growth and development—while creating environment *bads* initially—will eventually lead to a turning point where the environmental conditions begin to improve. In fact, there is the hope that at this juncture, environmental harm and economic growth would then become decoupled from one another.

Research on an environmental Kuznets curve has shown that such a model is flawed, especially from a regional or global vantage point.[30] For example, Mazzantu and Zoboli, examining packaging waste in Europe, do not see a turning point in the near future. Instead, they find that waste generation is increasing despite increased income. Moreover, they find that the environmental costs associated with waste have increased.[31] Similarly, Gnonlonfin, Kocoglu, and Peridy examine the greater Mediterranean region, which includes Southern Europe as well as countries in North Africa and the Middle East. They find the ability to reach a Kuznets tipping point only exists for wealthy countries. This is due to uneven development, in which less developed countries take on a more significant burden of handling waste within global trade networks of trash.[32]

While the crises involving garbage barges reveal the global nature of waste crises, the commodification of trash and its integration into the calculations of urban policymakers have significantly pushed urban ecological footprints to overlap more and more. This means that we need examine how waste management as a sector operates alongside global urbanization. Driven by financialization, the handling of trash is no longer an afterthought for a city in dire need of landfill space. In other words, it is not just more people resulting in more trash. Instead, trash is now a force that drives expansion and, in turn, creates new threats to cities.

Simply put, expansion leads to overlapping ecological footprints. Cities can often have several hundred times the ecological footprint compared to geographic area.[33] After all, cities (as well as nation-states) do not operate independently. They draw on resources from beyond their borders and export wastes beyond city limits. This includes rapidly urbanizing cities in the South. However, ecological footprints are not distributed evenly, and their impact is often externalized—to other cities, regions, and nations.[34] Thus, when we speak of cities having a shared ecological footprint and vulnerability, we also need to rethink the way cities operate globally. Rather than fixed points on a map, cities are part of constellations that cross regional and national boundaries.

Today, solid waste management is a part of city-building locally and globally. Urban planning and construction are multifaceted projects involving the construction of buildings and infrastructure development. Correspondingly, multisectoral approaches to urban planning are an ideal or a best practice in the building of cities. Traditionally, this meant coordinating several actors (e.g., architects, designers, engineers, utilities, property owners) to develop and construct the built environment. However, in the age of global neoliberalism, we see that large multinational corporations play an influential role in the construction of cities and their infrastructure. Often large holding companies own multinational subsidiaries that specialize in the work of city-building, which includes building and infrastructure construction as well as so-called environmental contracting or management.

The rise of neoliberal urban governance has allowed for the corporatization, financialization, and securitization of public goods such as MSW management. Here, the "commons" are privatized. Yet, deregulation does not mean the absence of regulation. Governments continue to enact law that govern these processes and neoliberalizing practices. Such policies give private firms reign over increasingly overlapping sectors such as waste, energy, water, and other infrastructure that provides public goods.[35] Politicians argue that such privatization "places a premium on efficiency."[36] While this is a prominent argument in countries such as the United States, we see that neoliberalism has embraced "green" urban development in the Global South as well. The push to modernize MSW supports economic and urban expansion—more so than sustainable development. As such, growth-oriented neoliberal policies "hijack sustainability."[37]

Take, for instance, the firm Actividades de Construcción y Servicios, S.A. (ACS). Headquartered in Madrid, the company specializes in construction and civil engineering services. It operates globally with half its revenue coming from the Americas and about a third coming from Asia.[38] It has, for quite some time now, moved beyond simple construction to environmental contracting. Environmental contracting includes so-called environmental services such as waste, energy, and related logistics. What this means is that ACS and its subsidiaries provide technical and logistical support at all levels of city-building, as well as the operations and management of municipal services. One subsidiary is international waste management firm Urbaser that operates in Spain, France, Greece, Italy, Portugal, and the United Kingdom. It is the seventh largest metropolitan solid waste company in Europe.[39] In 2016, Urbaser was sold to Firion Investments, a holding company indirectly controlled by Chinese companies Huayu Construction, Jiangsu Dagang, and China Tianying Inc. (see figure 3.1). Before the sale, Urbaser was a partner in Urbaser-Danner—another holding company—shared with Tennessee-based Danner Company in the United States. Urbaser-Danner owned KDM Empresas group, which runs Chilean waste management firm Starco Demarco. Starco Demarco is one of the largest providers in Chile, alongside Cosemar, Dimensión, and Gestión

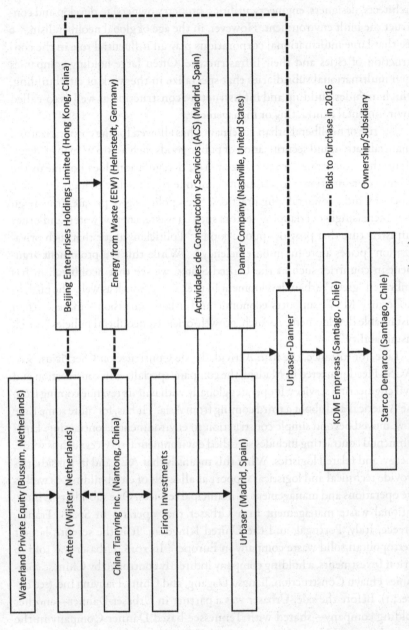

FIG. 3.1 Corporate connections in global metropolitan waste management (2016)

Ambiente. While Starco Demarco handles waste throughout the country, it has had a significant presence in Santiago's municipal areas since the 1990s.

The fact that Spanish, American, and Chinese corporations have owned one of Chile's largest waste companies reveals an interesting yet oft-told story of transnational corporations expanding into the Global South. Scholars have pointed to the automobile, pharmaceutical, and steel industries as drivers of internationalization. We also know that transnational real estate developers play a central role in global urbanization.[40] Underexamined, however, is the role of multisectoral megafirms that are actively involved in the physical development of cities (e.g., construction, infrastructure, and municipal services such as waste management). If one looks at the relationship between waste management and urbanization, we see that such megafirms reveal how urban metabolisms are global and interconnected.

The growth of private MSW firms within a major metropolis—such as capital or primate cities—subsequently gives these firms advantages for regional and later international expansion. Such firms will have accumulated financial and technical assets that allow them to win bids for MSW services in other cities. As these firms expand, it means a MSW network between cities takes shape. Cities within this constellation are encouraged to grow with greater reliance on each other for infrastructure development, services, and management of resources.[41] Put another way, firms that provide MSW services encourage urban growth that generates more consumption, which also makes cities more dependent on them. In turn, the ecological footprint of cities within this network increasingly overlaps and increasingly shares vulnerabilities—such as the collective need to manage and dispose of waste.

China's refusal to import waste is the best example of this changing global landscape of waste management. The country is not walking away from handling the world's waste. It is moving away from importing waste and toward managing waste.[42] The expansion of China's waste management interests began before China's National Sword policy banning certain types of waste from being imported. It is also tied to the Belt and Road Initiative by the Chinese government, which focuses on infrastructure development overseas. The initiative is most known for investing in transportation, energy, and trade networks. However, Chinese companies have begun developing waste infrastructure around the world over the past few years. It is easy to point to this expansion as evidence of China's influence on international markets or their economic agenda. Owing to the global nature of international finance, the expansion of Chinese firms is not unlike the growth of Western multinational corporations seeking to monopolize industries. Strategic acquisitions have given China a strong foothold in the world's waste management sector and especially the growing WTE industry.

China's foreign policy regarding trash and infrastructure is linked to urban growth. The rise of Chinese megacities also means that Chinese firms subsequently have accumulated financial and technical assets amid this expansion. This growth gives them the resources to either expand into other markets through the Belt and Road Initiative or buy out MSW services in other cities. This includes urban growth and developing MSW services that allow for more growth and consumption. The result is a constellation where Chinese, Spanish, and Latin American urban centers become more reliant on each other for ongoing growth.

Before the Urbaser sale, China Tianying Inc. made a failed bid for German waste management company Energy from Waste (EEW) in 2016. They also had an outstanding bid for Dutch waste management group Attero. Already mentioned earlier is WM's stake in SEG waste management–related activities. SEG itself is a subsidiary of Shanghai Chengtou Holding Co., Ltd.—a real estate developer that is building the Shanghai Tower, which is expected to be China's tallest building. These megamergers illustrate an urban growth machine dominated by multisectoral transnational corporations that not only deal in real estate but infrastructure such as waste management. Here, we can say that the mutual interdependence of cities under global capitalism leads to a shared ecological footprint in which their vulnerabilities are linked. Consequently, the analysis of such interconnections in waste management helps us understand the relationship between contemporary global capitalism and urban risk.

The rise of multinational corporations—especially those in China—is less surprising than the vertical and horizontal scale of corporate connections. Displayed on a map, this would appear to be a constellation of cities that illustrates contemporary global urbanization processes. We also see this in cities in the Global South. Based in Pretoria, The Waste Group began by handling local industrial and commercial waste in South Africa. It has now moved on to landfill management and even the recycling of construction rubble and debris. The company has now expanded into Mozambique, Namibia, and Ghana.[43] Companies such as The Waste Group are following in the footsteps of firms in the North and those in newly industrialized and semi-peripheral nations. Elsewhere in the world, firms such as Albayrak, Özkartallar, and Cam Pak have expanded throughout the country of Turkey and eventually overseas into NICs such as Pakistan.[44]

Logistics and Waste Management

The problem, however, is that globally, we produce more and more trash. This means that the threat of garbage crises is not limited to a single city or country. Garbage, instead, poses systemic risk. Capitalism has a solution—at least for the time being. It is the development and expansion of the logistics sector.

The management of waste involves the movement of material and, as such, is ultimately about logistics. Historically, logistics has been the management of resources as they move from the point of production to consumption. Increasingly, the resource flows managed by such companies include personnel, expertise, and other services. Thus, they have contracts to BOO or BOT a variety of different facilities and infrastructure. In the 1990s, logistics service providers began linking real estate development, construction, and property management services. This evolved from the need to move materials on and off construction sites.[45] Over time, companies came to provide additional labor and resources on-site and throughout the entire supply chain. This includes resource management and the maintenance of facilities such as airports, ports, and warehouses. Given the diverse range of roles that logistics service providers play, they are integral within large holding groups such as ACS.

In this sense, firms such as ACS are not just construction management companies. The Clugston Group in the United Kingdom, ORTEC in the Netherlands, GSE Group in France, Albayrak Group in Turkey, and many others are all as much logistics service providers as they are builders or developers. They are involved in real estate, construction, and the management of building site activities, such as bringing in goods (e.g., materials and tools) and removing wastes (e.g., scrap material and rubble) to allow for the efficient production of buildings. They also provide environmental and building management services after buildings are constructed. In a way, it is impossible to classify these companies beyond simply calling them multinational corporations.

The growth of multisectoral logistics service providers stems, in part, from the neoliberal argument that private companies are more efficient than state entities.[46] It is claimed that this is about the modernization of environmental services.[47] Mimicking ecological processes that channel waste products back into a productive ecosystem, this process has facilitated greater productivity in spite of higher demand for natural resources.[48] This framing is most evident in the cities of newly industrialized countries where "modernization" is a political objective.[49]

From a critical perspective, the modernization of garbage collection around the world is hardly sustainable or equitable.[50] Such firms are primarily driven by profit, not environmental concern. Cities in the Global South are catching up with the North in their own privatization of domestic public goods and the selling of services overseas.[51] Take, for example, the role of Turkish firms expanding overseas and exporting "environmental technologies."[52] In 2011, the Lahore Waste Management Company (LWMC) was created to modernize the city's infrastructure by rehabilitating landfills and creating new garbage collection infrastructure. From its inception, LWMC has worked with its counterpart in Istanbul (İSTAÇ), as well as Turkish firms Özpak and the Albayrak Group. These Turkish firms offered logistical support by providing trucks, personnel training, and constructing facilities. This partnership has since grown into other contracts with Albayrak and Istanbul Ulaşım Inc. to develop public transit systems. In addition to companies working in Pakistan, Turkish firms are also active in Eastern Europe and Africa. Turkish firm Evciler, which began working with the reclamation of chemicals and precious metals in the 1980s, has since moved on to e-waste and recycling both locally and abroad. Like Western nations, Turkey is driven by urban growth at home and expansion into overseas markets.[53] As cities grow and rely on such firms, they also begin to look regionally and globally for expansion.

Several perspectives from environmental and urban sociology, as well as world-systems theory, offer insight into this expansion. This convergence of large firms connecting urban ecological footprints can be understood through two processes well discussed in the environmental and urban sociology literatures: the "treadmill of production" and the city as "growth machine." Developed by Allan Schnaiberg, the treadmill of production is a model in environmental sociology that explains the insatiable and growing hunger for material goods. This growth-logic permeates society—economically, politically, and culturally. Social actors such as the state, the private sector, and organized labor all buy into growth and consumption, accelerating the treadmill of production.[54] A major fixture of this approach is the role of monopolistic firms. As Schnaiberg writes, "The volume and source of a treadmill of production is high-energy monopoly-capital industry."[55]

Large holding companies with many subsidiaries undoubtedly want to accelerate wealth accumulation. Expansion to other cities is a major way waste management firms accomplish this goal. Companies need cities and public agencies for their territorial expansion and legitimacy. Increasingly, policymakers operating under the guise of fiscal conservativism cede authority to them. In turn, firms increasingly gain power over local and national governments.[56] Again, waste *management* (which includes disposal, recycling, and waste-to-energy) should not be thought of as conservation. Not only are logistical services needed to move waste materials, but trash is commodified.[57] Rather, the handling of MSW should be thought of as *supply chain management*, which seeks capital accumulation, urban growth, the acceleration of the treadmill of production, as well as the redistribution of risk.[58]

Logistical services in waste management aid urban growth.[59] Logistics does more than connect urban areas. It makes cities mutually interdependent within the global economy or capitalist world-system. Logistics is a coping and modernization mechanism for the increasing speed and vulnerabilities of the global capitalist treadmill. Notably, the drive for profit promotes economic expansion, advances in technology, and consumption. This undermines improvements in waste management because consumption is not reduced. Here, the ever-increasing speed of the treadmill of production creates more waste and environmental degradation, which also has the potential of creating crises. However, logistical advancements in handling waste—locally and globally—temporarily lubricates the treadmill and allows it to run faster. It also does so by externalizing risk, by shifting hazardous materials onto global markets.[60]

In many cases, this means encouraging and facilitating urban growth regionally or overseas. As such, cities influence one another and create an ever-growing shared ecological footprint. As Frederick Buttel pointed out, the treadmill of production and the growth machine "depict powerful institutional pressures towards expansion and environmental degradation from the local to the global levels."[61] This is not unlike what John Logan and Harvey Molotch have termed the "urban growth machine," where business interests play a powerful role in urban expansion. Growth machines are traditionally comprised of developers, economic interests, and other social interests that benefit directly or indirectly from population increases and business expansion.[62] The growth machine is not without challenges. Cities are relatively fixed in space, with many cities having physical, geographic, as well as legal limits to their expansion.[63] This limits growth but also creates vulnerabilities, such as running out of resources or space for landfills. In addition, environmental movements—such as antigrowth or smart-growth initiatives—may fight against the machine.[64]

However, workers as well as the state buy into the growth-logic. While smart-growth, slow-growth, no-growth, and de-growth arguments exist, in the absence of successful social movements, economic expansion tends to win over citizens as well as organized labor. This drives the treadmill to run faster. The treadmill in turn accelerates the urban growth machine. Correspondingly, cities need to deal with their spatial fixity as well as vulnerabilities as they grow or else face disruptions in capital accumulation or resistance. Regional and overseas expansion, in turn, becomes important.

From the standpoint of world-systems, this makes sense. In the world-systems perspective, capitalists in wealthy nations look to expand markets, find cheap labor, and exploit resources. As such, core countries benefit from uneven development by taking advantage of less developed countries. This means the world economy operates as a singular capitalist system. Drawing on world-systems, Stephen Bunker argues that uneven development is a crucial component to the treadmill of production. The development of transportation and logistics facilitates not only the treadmill but the ability for capital to expand.[65] Similarly, urban growth machines—like countries—in this system can only grow if they go overseas. This not only creates new markets but also assists in the handling of waste.

As James Rice importantly points out, unequal sink-capacity between countries plays a significant role in the functioning of global capitalism.[66] A country's or city's ability to absorb environmental harm is very uneven. However, coping with sink problems is essential to the day-to-day operations of capitalism. While waste can be commodity, it can also be a hazard. Cities are, after all, nodes within the world economy. In turn, they are logistics centers and a space of flows. Large cities are also command points that direct materials from one city to another. In turn, the externalization of waste disposal is a product and consequence of an accelerating treadmill within the world economy that requires risks to be redistributed.

In today's system of cities, there are megacities with over ten million people, highly networked global cities, and rapidly urbanizing countries in the Global South. The urban treadmill is running faster than ever on a global level. Unsurprisingly, multisectoral city-building is increasingly in the hands of extremely large holding companies that operate globally and integrate MSW services and other forms of infrastructure development. Part of growth coalitions, these firms help mitigate some of the concerns regarding expansion while generating revenue from urban development

Incineration and Disaster Capitalism

Logistics commodifying trash in a consumption-driven feedback loop is perhaps most evident in incineration or, to use the industry term, waste-to-energy. WTE is celebrated in the trade and popular press, where it is touted as sustainable with phrases such as "energy recovery."[67] MSW providers also build incinerators to profit from the collection of trash. Aware of the stigma against incineration, press releases note the use of scrubbers and other technology that remove toxins. As such, incineration is promoted using terms such as "environmental responsibility" to assuage fears of chemicals harming human health. Corporations also celebrate landfill reduction and methane reduction in the atmosphere.[68] The problem is that neoliberal policymakers buy into this greenwashing of incinerators. In Arizona, for example, state utility regulators have argued that garbage incineration is a renewable energy source. This has been supported by local energy producers such as the Mohave Electric Cooperative and the Reclamation Power Group.[69]

Waste generation and energy consumption, after all, are expected to increase as the world's population urbanizes. In fact, the International Energy Association (IEA) noted in 2019 that there was a lot of potential for WTE in Asian countries as a result of urbanization, MSW generation, and economic growth.[70] Correspondingly, China has seen WTE as a means of coping with its solid waste problems. The purchase of Western companies with expertise in WTE, in other words, satisfies both domestic and foreign policy. In the past few years, China has built massive incinerators to deal with tons of municipal waste generated each year by its megacities. This, in conjunction with the purchase of foreign companies, allows Chinese firms to grow and dominate markets overseas.

Another way that WTE operators sell themselves as green is the marketing of incinerator bottom ash as a building material either in granule form or bonded to another material. The Confederation of European Waste-to-Energy Plants (CEWEP) has celebrated bottom ash a building material citing its use in aggregate for use in roads, foundation, and fill.[71] There is no doubt that ash generated from incineration can be and has been used in cities.[72] Correspondingly, WTE operators often qualify for green bonds that support both renewable energy development and transportation and recycling/waste management projects. Green bonds are a means of raising money for climate-related or environmentally friendly projects. While green bonds may encourage environmentally friendly or resilient urbanism, the underlying logic remains the financial returns on investments and reducing financial risk.

WTE is not actually green. Rather, WTE is a prime example of the treadmill of production and the role of logistics. The use of incinerators radically expanded the hunger for trash as it is the fuel used in the process. Since cities pay firms by weight, the more people consume and dispose things the more wealth is generated. We should not forget that these firms are often also builders of homes and infrastructure. This, in turn, increases consumption and waste. Logistics divisions then help coordinate garbage collection and incineration. In other words, these firms create waste, the logistical infrastructure to move waste, and transform garbage into a commodity through incineration. Thus, it has been argued that recycling does little to slow the treadmill.[73]

Simply put, corporations transformed "waste" into a source of profit by integrating it into their supply chains.[74] The use of incinerator bottom ash is less recycling and more reverse logistics. According to U.N. Comtrade data, the top importing countries of ash and residues from the incineration of municipal waste by weight in 2015 are Belgium, Qatar, Demark, and Germany. The trade value of the material was USD $14.7 million. While that value is quite low when discussing international trade, it is expected to increase as countries shut down their coal power plants—as Belgium did in 2016. Correspondingly, ash for cement increasingly comes from incineration rather than thermal power plants.[75] Here, we see trade relationships forming because of the reliance on WTE and ash. Sweden, for instance, incinerates Norwegian waste and in return gets incinerator ash. U.N. Comtrade data, again from 2015, showed that Sweden was the world's top importer of municipal waste, taking in 783 million kilograms of waste, with most of it going into incineration.

It is important to note that the foundation of modern cities is literally cement and concrete, whose production is a significant source of global carbon dioxide production. The generation of bottom ash from WTE helps in the production of cement for city-building, and ash is used in conjunction with traditional cement and concrete materials. Moreover, concrete, one of the world's most consumed substances, generates approximately 7 percent of all global carbon emissions.[76] As such, the so-called recycling of waste materials into building materials encourages urban growth and consumption.

Unsurprisingly, it is not just construction companies getting into the logistics of WTE. Major WTE players such as Covanta, Hitachi Zosen Inova AG, and Wheelabrator Technologies operate cross-sectorally. In fact, the second largest metropolitan waste management firm in Europe is Suez GDF/Engie. Suez is the same Suez SA that later merged with Gaz de France (GDF) to create Engie, one of the largest energy companies in the world operating in twenty-seven countries with a stake in power stations in the United States and the United Kingdom. GDF Suez/Engie stated that its goal is to provide "circular economy solutions . . . with the objectives of creating new resources."[77] In other words, the company follows a strategy of ecological modernization—seeking to solve the problems it creates. In turn, Engie has a bottom ash treatment facility in Grimbergen, Belgium, that handles 120,000 tons/year with plans to build more.[78]

In 2014, Suez collaborated with Beijing Enterprise to develop energy projects that include waste incineration in China.[79] Beijing Enterprise subsequently acquired, in 2016, Energy from Waste, which ran eighteen plants in Germany. A 2015 corporate statement from Engie noted that Brazil, Chile, Morocco, and South Africa were priority projects.[80] Suez is not alone with its sights on Africa. Currently underway is the construction of the Repi WTE plant in Ethiopia. This project is aided by UK-based Cambridge Industries and China National Electric Engineering Corporation. Elsewhere, Hygiene and Sanitation Company of Cameroon (HYSACAM) has taken on efforts to develop WTE plants with assistance from Spanish PIA Energy JVC Group.[81]

While these large corporations dominate MSW management, we need to remember that the development of services such as garbage removal and WTE is subsidized by the public sector. Cities are the foundation of this system. Municipalities providing waste management not only issue contracts for private companies to build or operate facilities, but they are also externalizing costs and risks. In turn, this aids the expansion of companies both regionally and internationally and, in turn, drives urban growth.[82]

The destructive tendencies of capitalism are clearly visible in how resources are withdrawn from the environment and additions such as waste are put back into the ecosystem. The growth of cities intensifies the withdrawal process. However, it is the development of the logistics/MSW sector that speeds it up by efficiently managing inputs and outputs. Thus, logistics accelerates the potential for capital accumulation. In other words, this sector allows the machinery of the treadmill to run faster through the efficient handling of "goods" and the management of "bads." By looking at logistics and waste management, the materiality of economic relationships is emphasized—which in turns reveals the shared environmental vulnerabilities of cities within the world economy.

Ecological modernization, however, does little to alleviate the destructive contradiction between society and environment.[83] This is in part the result of practices such as WTE serving elite interests and political power. While the focus of this chapter—thus far—has been on waste as a vulnerability, there are social vulnerabilities as well. As Schnaiberg notes, the treadmill of production encourages monopolies that undermine the sustainability of the social sphere as well. The growth of these holding companies and multinational waste management firms has consequences for workers. In the United States, companies such as Waste Management have been criticized for their racist and unfair labor practices.[84] In Chicago, this situation led to a 2003 strike by waste collectors. The expansion of Turkish firms into Pakistan led to strikes in Faisalabad and Ludhiana in 2013.[85] In 2013, there was a massive strike against Urbaser in Madrid.[86] In Chile, there were strikes among Starco Demarco workers in the summer of 2013 as well.[87] In 2017, protestors in the United Kingdom attempted to block the construction of an incinerator to be built by French Firm CNIM primarily for labor-related (rather than environmental) reasons.[88] In turn, garbage collection and politics are deeply intertwined, which makes it a site for collective action.[89]

Connected Vulnerabilities

As a result of the massive growth of these environmental management firms—linked to other sectors such as construction, energy, and urban development—a consequence is that local economic shocks can have systemic consequences. Natural disaster leads to infrastructure breakdowns and creates difficulty in disposing of post-disaster debris. Cities and government agencies find themselves coping with increased waste materials following major catastrophes. It is estimated that the cost of debris removal accounts for at least 27 percent of total disaster costs.[90] This is because some of that material is toxic. For instance, the United States Environmental Protection Agency has guidelines for dealing with disaster debris. Similarly, the United Nations Joint Environmental Unit and the World Health Organization have guidelines as well. Ultimately, the handling of post-disaster waste and debris involves many different actors in extremely strained circumstances.[91]

For example, the California Department of Resources Recycling and Recovery (CalRecycle) has to manage private contractors working to clean up disaster sites while also working with the U.S. Army Corps of Engineers to remove ash, remnants of burnt buildings, and other destroyed material, and deal with hazardous materials created by fires. In the case of Santa Rosa, California, an estimated 200,000 tons of material were collected in the month after the October 2017 Nuns and Tubbs wildfires. Waste collection continued well into 2018. In the ensuing months, nearby landfills were facing problems handling all the waste. The Army Corps of Engineers reports that a total of 2.2 million tons of debris, or the equivalent of two Golden Gate bridges, were removed in the fire season.[92] California is not alone. Other areas, such as Sevier County, Tennessee, and Fort McMurray, Canada, have had problems with postfire waste materials because of landfills having to process non-disaster waste along with all the additional debris coming in.[93]

Of course, the private sector has taken notice. The expansion of the restoration industry—made up of companies that are involved in a wide range of disaster-recovery and restoration services—is a prime example. Like the aforementioned logistics firms, they provide engineering and construction services as well as waste management following disaster events. While many firms help individuals fix their homes after a disaster, they are increasingly working directly with insurance companies and government agencies to clean up disasters. For example, the Fluor Corporation has had an exclusive contract with the Federal Emergency Management Agency (FEMA) since 1997 and has been on-site at the World Trade Center, in New Orleans after Hurricane Katrina, as well as in Florida following recent hurricanes.[94]

In conclusion, this chapter does not provide a complete survey of the interconnection between cities, waste management, and logistics in urbanization. However, it is a look at the relationship between MSW management and urbanization. In turn, we see that these connections reveal how urban metabolisms are increasingly interconnected. Cities are not only producers of waste; they are where logistical resources are concentrated. As such, the study of waste management and the logistics sector can reveal the "nature" of economic expansion.

4

Fire, the Wildland–Urban Interface, and Feedback Loops

• •

> All things are an exchange for Fire, and Fire for all things, even as wares for gold and gold for wares.
> —Heraclitus of Ephesus

This chapter examines the social, economic, and political relationship between urban development and fire. Wildfires (also referred to as bushfires) are increasing around the world because of climate change and urban growth. Wildfires in the past were largely local or regional events. Yet global climate change and human activities have made them a global problem. In turn, Andrew Scott describes our planet as literally burning. He and others note that wildfire is not just a natural disaster: It is a human and urban catastrophe that is happening on a larger and larger scale.[1] This has profound consequences for our understanding of urban nature.

Globally, an estimated 300,000 deaths and millions more injuries are caused by fire each year.[2] Fire also destroys buildings, infrastructure, as well as threatens wildlife. Mediterranean climatic regions—Australia, California, Chile, Mexico, Southern Europe, and the Mediterranean Basin, as well as South Africa—have seen an increase in fire intensity. In February 2017, nearly 100 fires raged across Australia's New South Wales, a region that regularly sees massive bushfires. These fires have been compared to the 2009 "Black Saturday" blaze in Victoria that killed 173 people. The 2019 to 2020 Australian bushfire season, which burned 18 million hectares, has come to be called the "Black Summer." In 2017, California broke numerous wildfire records. That year, five of the worst wildfires in the state's history occurred, making it the most expensive year of firefighting at the time. In 2018, the Attica Fire in Greece killed ninety-six people, making it the second deadliest wildfire of the century, after the Black Saturday fires. In 2021, Turkey saw fire threaten a coal power plant and burn 170,000 hectares.

In addition to Mediterranean climatic regions, in 2015, over a million hectares of boreal forest in Siberia burned. In 2019, about 13.1 million hectares were decimated by forest fires, resulting in the declaration of a state of emergency. The 2016 Fort McMurray Fire in Alberta, Canada, was one of the costliest fires in Canadian history at the time. Approximately 600,000 hectares were burned and 2,400 structures were destroyed. Included in the cost is the disruption of Canada's oil production.[3] Throughout summer of 2018, Sweden, Latvia, and Finland had to battle an unusually high number of wildfires. The world, simply put, is on fire.

The threat of massive wildfire is driven by urban growth, so-called economic development, and climate change. Larger populations and more infrastructure are exposed to hotter and drier conditions, which creates an increased risk of disaster.[4] There is no doubt that urban development, which pushes human settlements into close proximity to flammable vegetation, has played a significant role in increasing the intensity of massive fires. For instance, the wildland–urban interface—where settlements and vegetation meet to create a combustible environment—are particularly vulnerable areas for wildfire.[5]

This relationship between fire and urban growth is tied to human negligence, including arson, campfires, cigarettes, electrical equipment failure, fireworks, and waste incineration. As discussed in chapter 3, the production of garbage, exacerbated by urban growth, makes not only for unsustainable cities but also for *risky cities*. In fact, it is estimated that between 15 percent and 25 percent of wildfire cases around the world are caused by the burning of trash. The burning of trash left by tourists along the Mediterranean coast and camping in the United States leads to wildfires, as does the incineration of trash in open-pit waste dumps in developing countries.[6]

The cause and consequence of wildfire is social. Whether it is human causes for fire ignition or the loss of human life and property, fire is a social disaster. Humans change the physical environment by building homes, cutting down trees, changing waterways, and creating a drought. They can also bring invasive species such as cheatgrass (*Bromus tectorum*), which can serve as fuel, into an environment. Urban expansion often occurs in bush, chaparral, forests, and other ecologies that are particularly prone to wildfire.[7] Human settlements in the wildland–urban interface bring people closer to combustible material. Then human action ignites the landscape, creating a disaster. Ongoing construction in these areas, in turn, creates a dangerous feedback loop (see figure 4.1). This means that fire or wildfire risk is a by-product of social and economic forces that drive urbanization.

We also see that modern capitalist society has commodified fire and risk in new ways, turning the environmental *bad* into a potential *good*. When I say that it becomes a *good*, I do not mean it is beneficial for all people or the environment. In fact, disaster's effects and consequences are different for different people. Fire does not hurt the property values of wealthy enclaves constructed in vulnerable ecologies. The 2003 Cedar and 2007 Witch Creek fires in San Diego, California, had more significant impact on middle and low-income zip codes than extremely wealthy ones. In fact, elite gated communities continued to be developed in the scorched neighborhood of Rancho Bernardo in San Diego. In the case of Australia, population levels and growth rates generally return to pre-disaster levels in a few years. The Kinglake region affected by the 2009 Black Saturday fires had recovered its population and gained new residents by 2014.[8] This not only illustrates how disaster does little to impede urban growth and development, but it reveals how powerful real estate interests are in disaster capitalism.

Fire-related disaster capitalism is linked to the way in which capitalism increases risk, while elites profit from risky urban expansion. This chapter examines wildfire in relation to urban growth in two ways. First, this chapter examines the problem of urban growth. While cities have always been affected by fire, global climate change and urban growth have increased wildfire risk. Global warming has exacerbated fire risk by creating drier, drought-prone regions, which in turn creates fuel for longer fire seasons. As humans build into the wildland–urban interface, new communities are exposed to wildfire risk.

Hundreds of thousands of people around the world are affected by wildfire. Like other so-called natural disasters, the damage and harm created by wildfire is a by-product of capitalism's growth-logic. Ted Steinberg argues that the typical political response to natural disaster is one of inevitability or that the catastrophe is an "Act of God."[9] However, this is not the case. Humans make voluntary or forced decisions to build homes and live in fire-prone wildland–urban interfaces. The building of communities and cities in such ecologies is driven by the capitalistic drive for growth. This drive for growth not only encourages urban development in the wildland–urban interface but also encourages rural-urban migration, which influences a region's fire regime. As populations shift from the countryside to urban and peri-urban regions, we see that previous patterns of land use that reduce wildfire risk, including traditional farming and brush-management techniques, are lost.[10] As such, urbanism must be examined within a broader ecological context.

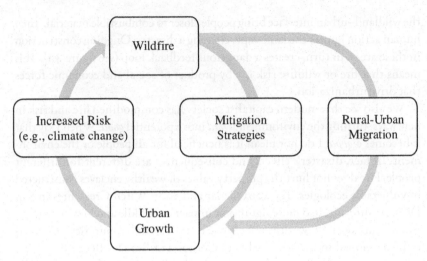

FIG. 4.1 Wildfire and urban growth loop

Simply put, an increased concentration of people living closer to flammable material and increased temperature makes for a risky situation. The risk-management company CoreLogic estimates that 869,102 homes in the United States are in high-risk areas. This is a significant increase from an estimated 177,000 homes within such regions in the 1990s.[11] According to the Department of Environment, Land, Water, and Planning in Australia, Melbourne's peri-urban expansion, the construction of holiday homes, and poor land use have exacerbated fire risk.[12]

This is particularly risky in low-income countries with informal settlements. In such cases, fire can destroy whole communities. South Africa, for instance, not only deals with regular wildfire but people living in informal settlements. These places are particularly vulnerable because of crowded flammable structures and lack of coordinated fire protection and fighting. On New Year's Day 2013, the township of Khayelitsha on the periphery of Cape Town saw nearly 4,000 people displaced. The township has since experienced several more shack fires killing people and destroying homes.[13]

The 2014 Valparaíso Fire in Chile began as a forest fire and progressed through the wildland–urban interface, affecting large sections of the city. The result was fifteen people dead, hundreds injured, and thousands of homes destroyed. During the fire there were reports of miscommunication and jurisdictional conflicts that hindered firefighting efforts owing to the scale of the conflagration.[14] Yet, we must remember that such fires are not singular events. Another fire hit Valparaíso in 2019, destroying 200 homes, indicating that fire is symptom of systemic social, political, and environmental problems.

As such, the second issue this chapter addresses are the problems of wildfire and disaster capitalism when building cities. Capitalists benefit from real estate development in fire-prone regions. Given the profit to be made through urban development, disaster mitigation or adaptation strategies are less about slowing growth than about reducing damage enough to encourage development. This includes not only real estate development but the construction of infrastructure. To facilitate this growth, insurance, new technologies, and private wildland-management contracts allow for ongoing sprawl deeper and deeper into the wildlands, foothills, and other fire-prone regions.

While policymakers generally recognize the social and economic dimensions of wildfire, the framing of solutions as "resource management" suggests that disaster mitigation can be profitable.[15] Moreover, it is clear that the alleviation of risk or harm is not without biases. Neoliberal land-use policy encourages or facilitates real estate interests—often at the expense of the general public.[16] Urban growth produces risk that forces cities to mitigate hazards, which encourages more development (see figure 4.1). The mitigation of hazard is not the same as adaptation to climate change or the development of more resilient infrastructure. Disaster preparedness, new technology, and better planning may still increase risk. Rather, mitigation is often a form of ecological modernization that allows for ongoing risky urban development. This feedback loop illustrates how disasters such as fire can shape risk for cities.

Climate change has profoundly changed agriculture and traditional livelihoods.[17] Climate change is one of several processes whereby humans have changed the environment, which includes draining water supplies, drying out the physical landscape, and bringing human negligence into a region. Urban growth fuels carbon dioxide (CO_2), which exacerbates droughts resulting from climate change. This generates food insecurity, which encourages rural–urban migration. In the developing world, this has a particularly devastating impact on human well-being. We also exist in a world where international food markets favor monoculture and cash crops. This has led to the decline of small farms.[18] The result is abandoned fields and grazing lands, which are at greater risk of wildfire because of their desertion.[19] Around the world, from developing nations to Europe, we see rural abandonment increasing the volume of flammable brush in already-vulnerable woodlands.[20] Also, severe wildfire can further drive rural–urban migration by destroying livelihoods in the countryside. This is a feedback loop whereby growing cities exacerbate the climate change that makes rural areas vulnerable to wildfire, which in turn contributes to ongoing urban growth.

History of Urban Fire and Growth

According to Stephen J. Pyne, "fire ecology is human ecology."[21] Both city and countryside are built environments that change over time because of human activities. In turn, so do their fire regimes—the historical trends and patterns of wildfires that prevail in a region. While cities are often thought of as the antithesis to nature, they rely on the countryside for natural resources and places to build.[22] This relationship between city and countryside is often one of conflict because of urban expansion and environmental degradation.

It is important to distinguish the role of contemporary capitalism in increasing fire risk from how cities lived with fire before. Cities have long lived with fire. Greg Bankoff notes that "preindustrial cities burn frequently, and on a scale rarely seen today."[23] Indeed, cities in the past dealt with fire regularly. We do not see the regularity of urban fire in major cities as we did in the past. However, the megafires we see today—killing people, burning thousands of acres of land, destroying property, and forcing evacuations—are now seasonal events of ever-increasing devastation.

At the same time, it is useful to look at the evolution of urban and wildland fire to understand how contemporary capitalist cities deal with disaster. By doing so, we can understand how cities evolved because of catastrophe and how risk is both a product and a consequence of the built environment. For instance, the proximity of dwellings to one another and the lack of modern firefighting made cities particularly combustible. Rome in 64 C.E., Amsterdam in 1421, and London in 1666 are examples of just a few infamous conflagrations that occurred before the Industrial Revolution. The haphazard layout of earlier cities was driven by urban growth and a lack of urban planning. This, combined with the lack of modern firefighting techniques, created a built environment where flames could leap from one building to the next and quickly destroy whole neighborhoods.

Fire was among the first so-called natural disaster for which cities developed mitigation strategies. Medieval cities, as Lewis Mumford chronicles, passed ordinances restricting roofing material and construction materials by the thirteenth century. For instance, the thirteenth-century fires of Bruges, London, and Lübeck led to the cities requiring the use of stone as a fire-resistant material.[24] Correspondingly, Daniel Turbeville argues that fire was a major driver of urban change well into the nineteenth century—transforming "cities of kindling" into more fire-resistant settlements."[25]

Even in early modern Europe, post-disaster cities were ripe with opportunity for rebuilding. Following the Great Fire of London in 1666, famed architect Christopher Wren laid out plans for the rebuilding of the city. Although his plan was not adopted, the intent was not unlike Daniel Burnham's plan following the Great Chicago Fire of 1871. Both Wren and Burnham sought to transform their city's street system, produce grand civic spaces, and essentially modernize the city. At the same time, such plans would make the city less combustible by rationalizing the city's plan.[26] Indeed, as Christian Pfister argues, disaster has historically created opportunities for society to modernize. People, after all, can learn from natural disaster. The question, however, is whether the desire for resilience overrides the interests of property owners.[27]

The modernization of urban services—especially in regard to fire—coincides with the development of various profit-seeking enterprises.[28] Take, for instance, firefighting. Marcus Lincinius Crassus created the first private fire brigade in Rome. He also used fire—including actively burning buildings—to buy up property. The profiteering was so apparent that historian Plutarch remarked that "if we may scandal him with a truth, he got by fire and rapine, making his advantages of the public calamities."[29] Later the *Vigiles Urbani* or "watchmen of the city" were established to fight fire as well as serve other civic functions, including enforcing general fire safety practices in the city. Following the Great Fire of London in 1666, Nicholas Barbon created the first private insurance company, the "Fire Office," which offered fire protection to clients. Insured buildings were affixed with a fire insurance mark that let an insurance company's fire brigade know whether it should be protected. It was not until the nineteenth century that professional municipal fire brigades emerged in the United Kingdom and the United States.

While large cities today are served by professional firefighters, firefighting historically has been done by volunteers around the world.[30] This is because fire was largely a local event that demanded immediate action from those present in the neighborhood. However, as cities grew—not just in population, but in physical form—cities created municipal fire agencies to coordinate firefighting and prevention activities. As Amy Greenberg argues, the status of a city's fire department was a part of late nineteenth-century boosterism. Professional firefighting, in this era, was used by so-called boosters to advertise a city as modern and promote urban development. As Greenberg points out, "A paid fire department was considered progressive, and more importantly, it promised a future of progress."[31]

It should be noted, however, that the literature on urban fire is typically examined separately from that on wildfire. This is due in large part to the urban–rural divide in various research areas. Work on urban fire is largely centered on city planning or disaster response. For instance, the historical work on the 1871 Chicago Fire and the fire following the 1906 San Francisco earthquake often examines haphazard city-building and the lack of effective firefighting strategies. Recent studies on shack fires in newly industrialized countries similarly look at how informality in the urban built environment creates precarity. Yet building code and planning in modern cities can be undermined by policymakers. Joe Flood suggested that the massive fires of the 1970s in New York City were fueled by budget cuts and technocratic decisionmaking—driven by questionable algorithms. In addition, Mike Davis noted the role of neoliberalism in the mishandling of fire in California, specifically citing the diversion of public monies away from public services.[32]

Work on wildfire, meanwhile, is largely based on forestry, land-use management, and environmental studies. Early research into wildfire, conducted as early as the 1930s, found that human activity was a major driver of wildland fire. For instance, data contemporaneously collected by the forestry service in Ohio between 1926 and 1935 indicated that the burning of debris caused 24.4 percent of fires, followed by cigarettes and arson as major causes of wildfire.[33] Recent historians examining the massive wildfire of 1910 that affected the states of Washington, Idaho, and Montana have looked at land use and the lack of a firefighting system to deal with such a conflagration.[34] Today, research continues to focus on human activity and its role in wildland fire around the world—for example, the activities of peasants and others living in rural areas that engage in pastoral burns, accidents in cooking, and other traditional activities that could ignite a fire.[35]

However, this division between urban and rural research has been criticized by contemporary scholars. Rural and urban regions are without doubt interdependent socially and economically. Whether it is rural-to-urban migration or the flow of resources, cities and the rural periphery are interconnected.[36] Cities also expand into rural areas. In turn, there is a body of work on the wildland–urban interface from the standpoint of urban planning, sustainability, and ecology. Urban and suburban development into wildland areas has caused environmental harm, such as reducing biodiversity and harming wildlife. However, it has also increased wildfire risk. As such, not only do cities have a shared ecological footprint—through infrastructure—but they also share risk with the rural countryside.[37]

Take, for instance, South Africa. The region around Cape Town has been particularly vulnerable. The fynbos in the Western and Eastern Cape of South Africa has plant species that germinate following a fire, not unlike the chaparral in California. At the same time, urban growth in this pyrophilous or fire-loving ecosystem has created an increased risk of fire. This situation is exacerbated by strong winds along hillsides that fan flames and make firefighting difficult.[38] In 2017 and 2018, several fires put coastal towns along the country's scenic Garden Route at risk and forced the evacuation of thousands. This is partially driven by a warmer and drier climate—leading to severe drought and an increased fire risk. At the same time, drought and reduced water supplies are worsened by consumption from growing cities. Throughout 2018, Cape Town faced a severe water crisis, leading to significant water-rationing measures. This had implications for firefighting, with the city's Fire and Rescue Services having to evaluate the sources of water it draws on to fight both bushfires and structural fires. As such, like many cities around the world, Cape Town's problems with fire reflect the convergence of climate change, ecology, and urban growth.

The Built Environment and Outsmarting Disaster

On the surface, constructing fire-resistant buildings and developing strategies for outsmarting disaster has benefits. Indeed, cities have become safer because of building codes and the use of fire-resistant architecture. The built environment appears to stand valiantly against nature thanks to ingenious solutions developed by human beings. For instance, architects in California have now accepted wildfire as a fixture of the landscape. This meant designing homes that interact with nature and implementing masonry and concrete that increase fire-resistance.[39] More recently, cities have developed mitigation measures such as brush clearance, controlled burns, and fire breaks to prevent fire from reaching residential areas. In other words, fire prevention and defense strategies are increasingly integrated. Homes are expected to be built from fire-resistant material, and residential fire codes now include a brush-free perimeter. This is to prevent fire from creeping up to the front door. Finally, we see new communications media technologies working in conjunction with warning systems to inform people of nearby threats and deliver evacuation orders.[40]

Such techniques for combating fire risk suggest a belief that safety is an achievable reality. They are technical fixes that assume that it is possible to decouple safety from urban development. This is partially due to the way in which some policymakers believe they can design away fire risk. It is a technocratic ideology that emphasizes technology and management solutions—especially those that reshape the built environment.[41] Such solutions are often shortsighted. The creation of fire-resistant materials such as cement and concrete are an example. As briefly mentioned in chapter 3, cement and concrete production contribute to global carbon emissions. While concrete production contributes to fire-promoting climate change, the material was touted in the early twentieth century as fire-resistant.[42] In turn, concrete became a staple of modern city-building and large-scale engineering projects. Yet its production requires the running of massive furnaces. Given the energy intensity that goes into its production, as well as the persistent and growing global demand for the material, its use is not necessarily making us safer.[43]

In turn, the strategies to deal with wildfire risk reveal an anthropocentric hubris that we can conquer nature. Like hard fixes mentioned in chapter 2, it is believed that we can engineer our way out of disaster. Yet, generally absent from some of the technical literature surrounding fire mitigation is the role of contemporary capitalism, politics, and social consequences. Or put another way, adapting human behavior and institutions to nature is secondary to adapting the physical and natural environment to economic needs.

There is no doubt that such strategies can mitigate or lessen impacts. Fire is not absent from the economic calculations of developers, financiers, insurance companies, and policymakers. On the contrary, fire is now part of the risk models of disaster capitalism. We see that such policies—while perhaps put into effect by public officials—are implemented through private contractors. Their deployment is typically in the form of managerial/technocratic practices influenced by neoliberal ideology. In turn, there is an endless drive to produce and sell new technologies and strategies for managing disaster risk.

Privatized disaster prevention and management are not value-neutral. While the disaster-management industry can protect lives and property, profit-driven entities do not have a strong interest in reducing overall risk. The disaster-industrial complex thrives on risk. Actors in this sector ultimately reflect the values of contemporary financialized capitalism and the urban growth machine. In turn, we see that the reliance on private builders, firefighters, insurers, and restoration services is directly linked to declining funds for public fire safety. These firms have found a market in a world that is warming, while cities sprawl throughout fire-prone geographies. Even if such technologies are properly implemented, the risk created by climate change and human settlements expanding into fire-prone regions is ignored.

Contemporary capitalism has placed a great deal of emphasis on outsmarting disaster through "innovation." The outcome of this approach has been the creation of defensible spaces. Private developers, currently design fire-resistant enclaves that are cared for and maintained by private landscapers. This is also seen in how private contractors are increasingly called upon to fight fires for the super-wealthy directly. Much like earlier models of insurance and the use of fire marks, these contractors often work on behalf of insurance companies, working to protect their clients located in high-value zip codes. Wealthy homeowners with AIG, Chubb, and Allianz policies may have mobile units deployed to their homes to spray them with fire retardant, as well as to provide customized services for evacuees. Specifically, companies such as Wildfire Defense System work with insurers to, in their words, provide "programs for asset protection and loss prevention."[44] This is true in the case both of enclaves in the wildland–urban interface and of large landowners who often want to avoid government agencies on their property.[45]

The creation of defensible space by wealthy homeowners is not coincidental. Elite enclaves are already designed to control space. Using these strategies to handle disaster is not unlike the way urban planning is used as a counterinsurgency measure—"protecting" wealthy neighborhoods by keeping unwanted people away. Just as architectural design may secure buildings from intruders, the built environment defends spaces from disaster and protects investments. In other words, defensible homes are a product of both the social and the political environment. They are a product of wealthy people building in desirable yet precarious landscapes. As such, we need to ask: Who can afford what kind of home in a particular place? Why are these places also considered valuable real estate? What is driving up property values, despite proximity to risk? What sort of strategies can be deployed to reduce risk, rather than just mitigate damage?

Regarding hard fixes, building codes and architectural fixes are, on paper, tactically sound. However, they do little to address the underlying social and economic dimensions of wildfire risk—urban growth and the rise of disaster capitalism. Adapting the physical built environment does not represent a substantial change to the core driver of wildfire risk—more people and buildings put on the frontlines of a firestorm. Take, for example, fire codes. There is no doubt that the use of fire-resistant material mitigates damage. As mentioned earlier, cities have regulated building construction to reduce fire risk. In 1872, the Chicago City Council mandated the use of fire-resistant materials. The Baltimore Fire of 1904 also resulted in new building codes, as did the 1911 Triangle Shirtwaist Company Fire in New York. In this period, organizations such as the National Fire Protection Association and the National Board of Fire Underwriters began promoting the use of iron, concrete, and steel in American cities.[46]

However, it is important to note that in the early twentieth century, there was a push back against building codes and ordinances. While insurance and underwriting bodies were in favor of building codes, those involved in construction and building were hesitant to support regulation. Public building officials in booming cities expressed concern regarding the use of "police power" in enforcing building code, ordinances, and zoning. Regulators gathered at the 1915 Building Officials Conference in Pittsburgh, Pennsylvania, warned that building code and zoning interfered with private property rights. Such officials, despite having the authority to regulate buildings, showed great deference to the construction and real estate sector. In 1925, following the California Supreme Court's decision authorizing cities to enforce building code, Gordon G. Whitnall, the city planning commissioner in Los Angeles, suggested that "it should be borne in mind that Zoning is but the term applied to the use of the police power as a means of publicly regulating the use to which private property may be put."[47] This response was undoubtedly due to public officials working with boosters and the construction industry to promote urban growth.

Nonetheless, building and fire codes became an accepted norm. In 2005 and 2007, the California State Building Standards Commission amended the California building code to provide stricter regulation in fire-prone wildland–urban interface areas of the state in response to earlier megafires. Effective beginning in 2008, revised code 701A.3.2 not only dealt with issues of landscaping but also required the use of fire-resistant materials in new structures—such as brick, cement, concrete, and tile. In 2013, the state of California passed Assembly Bill (AB) 1492, which placed a tax on wood shingles and other lumber products. This bill encouraged the use of fire-resistant roofing, but also used tax revenue to limit the lumber industry's liability for fire started on their land. In 2018, a bill was proposed in the State Assembly that would allow renovation costs for projects that increase fire safety—such as replacing wood shingle roofs—be used to offset property taxes.[48] These changes, however, were the result of disaster creating the political will to create new taxes and regulations. Yet tax revenues appear to support private property interests rather than broader fire risk reduction projects.

Again, building code does little to stop development and sprawl. Rather, the goal is to make the built environment more effective in its defense against fire. This is a form of shelter-in-place (SIP), which refers to the idea that one can create defensible or secure sites where evacuation is a last resort. It is a strategy that has been applied to chemical disasters as well as suggested for terrorist attacks. It is not unusual for emergency officials during a disaster, pandemic, or active-shooter event to request people to SIP or stay put until asked to evacuate or clear the area. Such plans, however, typically involve designating or creating defensible spaces such as postnuclear bunkers and safe rooms with air filtration in case of a biochemical attack. Shelter-in-place is appealing because it offers a technical solution to our problems.

Central to an SIP implementation is the calculation, planning, and management of risk. Officials and planners must calculate the defensibility of spaces versus the time needed for emergency evacuation. Again, SIP focuses on hard or physical solutions such as architecture and other technologies that heroically defend a structure and its inhabitants from danger. It also makes seemingly obvious solutions—such as using fire-resistant building materials and landscaping—seem ingenious. Moreover, by making evacuation a last resort, it exemplifies an anthropocentric hubris that believes that we can challenge or fight nature. In fact, an often-heard mantra of SIP is "stay and defend."[49]

As noted earlier, medieval cities, as with cities today, used architecture and planning to reduce the risk of fire or to make it easier to fight fire. However, SIP also includes a policy decision: ordering evacuation. When California policymakers were adopting the strategy in the early 2000s, they looked to Australia's "stay or go" or "stay, defend, or leave early," policies that were developed in the mid-1990s. Procedures developed in Australia included preparation, pre-emptive defense, and evacuation orders in the event of bushfires. The strategy was celebrated for its emphasis on defense, but it has since been absorbed into more systematic disaster-management plans. While Australian disaster management has influenced California, there are significant differences in policy, culture, and history between the two places. What makes southern California different is that there is a greater reliance on private entities to design and maintain safe communities. This includes the use of private contractors to manage brush and provide landscaping services as well as firefighting activities. Correspondingly, there is less emphasis on public fire prevention and suppression.[50]

California's approach seems to ignore the relationship between calamity and capitalist urban growth. Humans try to mitigate wildfire risk while encouraging more urban development, which again creates risk. The contradiction is at the center of disaster capitalism. As mentioned in chapter 1, it can be argued that disaster and crisis are deeply ingrained within the economic system. The problem lies in market-oriented innovation that encourages the real estate industry to sell risky development properties as if they were stocks rather than homes occupied by people.

Take, for instance, San Diego County, which saw massive fires in 2003 and 2007 yet lacked a centralized fire authority until 2008. Attempts in 2003, 2008, and 2009 to collect fire protection fees within State Responsibility Areas (SRA) to fund state-level fire protection were either repealed or withdrawn from the legislature. Finally, in 2011, the Fire Prevention Fee was enacted and went into effect in 2013. However, the collection of the fee was suspended in 2017 until 2031.[51] In turn, the Riverside County Fire Department's website notes that crews from the California Department of Forestry and Fire Protection (CalFire) dropped from 245 in the 1980s to 197 in the 2010s because of budget reductions. This means that "during serious fire sieges, 170 to 175 fire crews are committed to major fires leaving only about 20 crews available for new incidents."[52] This is in addition to the fact that CalFire exceeded its budget in years with particularly bad fires. For example, by mid-2017, CalFire had already exceeded its annual budget by $60 million. Similarly, calls for support at the federal level to fund the U.S. Forestry Service have not worked, either.[53] The U.S. Forest Service has had to rely on other federal agencies such as the Department of Homeland Security to cover the cost of fighting fire. In turn, California had to create special funds to deal with fire protection and provide insurance.[54] So, it follows that there are gaps in fire protection between different levels of government, which has made code enforcement and the construction of fire-safe buildings at the local level much more central to firefighting.[55]

In turn, fire risk has not gone away—or been reduced. The catastrophic 2017 and 2018 wildfires in California are part of a trend where wildfire size and damage have increased year after year (see figure 4.2).[56] Of the twenty largest fires in California's history, fourteen have been since 2000—a period that saw ongoing population growth.[57] In response, the state has changed fire codes and deployed a number of strategies to fight fire and mitigate damage. Indeed, a retrofitted house is more fire-resistant. However, fire resistance creates a false sense of security as it does not mean that the property is fireproof.

This false sense of security has had implications for fire-insurance coverage in the state. Following a series of catastrophic fires in California, the insurance industry has expressed interested in pulling out of the state. In turn, California

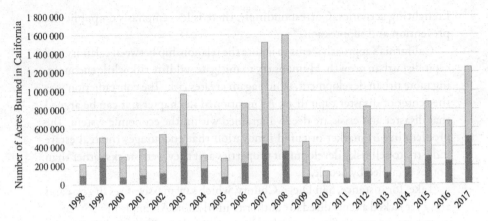

FIG. 4.2 Acreage burned in California by jurisdiction (1998–2017)

has expanded its state-administered fire insurance program. Lawmakers are also trying to develop ways to assist private insurers in providing coverage. Despite attempts to outsmart fire and mitigate financial costs, we see that corporations are ultimately concerned with profit margins. Without strong regulation of both the built environment and the insurance sector, risk is ultimately borne by people, communities, and cities.

Neoliberalism: Shifting Risks

Firefighting in the wildland–urban interface often involves coordination among various agencies—national forestry services, local firefighters, and private entities. In the United States, the National Cohesive Wildland Fire Management Strategy involves the U.S. Department of Agriculture, the Forest Service, the U.S. Department of the Interior, and state forestry agencies. Rangeland Fire Protection Associations—or state-authorized volunteer firefighters—provide fire control in the American West. Similarly, researchers argue that these initiatives are highly dependent on coordination among communities and different agencies.[58] Rather than the U.S. federal government playing a larger role in coordinating firefighting activities, we see local and state entities increasingly carrying the burden. After all, a by-product of neoliberalism is political aversion to public expenditure on disaster mitigation and relief, which includes funding coordination efforts.[59] That said, local, state, and federal agencies are doing the best they can, despite being under-resourced.

As Sorensen and colleagues point out, protective action is resource dependent.[60] Given the unequal distribution of resources, we must ask who fights for what. Are safety and disaster resilience merely cover for neoliberal policy? In the case of wildfire, fire prevention has inevitably become intertwined with a variety of different political efforts. For example, there are conflicts that exist between environmentalists, developers, indigenous populations, landowners, and government initiatives. Rural and indigenous populations in developing counties are at times in conflict with state agencies regarding traditional farming techniques. Sometimes, governments will restrict pasture burning. Yet restriction on traditional farming techniques can disrupt local communities that have historically kept flames at bay through their farming practices.[61]

To deal with declining U.S. federal support for firefighting, neoliberal policymakers have expanded or deployed several strategies to deal with fire. While neoliberal urban governance abdicates public responsibility, politicians deploy short-term "fixes" to society's need for public goods such as firefighting. As a result of massive wildfires in 2017 and 2018, there has been concern about firefighter shortages in California and other states. To recruit more volunteer firefighters, California Assembly Bill 2727 was proposed in 2018 to allow firefighters to claim a tax credit. Yet, it has been suggested that such a plan would cost the state $30 million in lost revenue—that could otherwise be used to fight fires directly. This, after all, does not solve the problem of recruiting volunteer firefighters and paying for fire suppression.[62]

There are even more serious consequences for such policies. As Loïc Wacquant has argued, neoliberalizing policies include the exploitation of incarcerated people to further market-oriented ideologies.[63] As a result of firefighter shortages, California has dramatically increased its reliance on inmates for wildland fire protection. It is estimated that 40 percent of California's firefighters are inmates taking part in the Conservation Camp Program that is overseen by the California Department of Corrections and Rehabilitation and CalFire. There are currently over 4,000 prisoners working in this capacity at forty-four camps throughout the state. Hundreds of low-level offenders were deployed in the 2014 Bully Fire in Shasta County. During the 2017 Thomas Fire in Santa Barbara, over 1,600 inmates worked on the fire line. In 2018, over 2,000 inmates, including youth offenders, worked twenty-four-hour shifts fighting the Mendocino Complex Fire. The vast majority are men of color and are paid $2 per day and have—according to some estimates—"saved" the state $80 million in "labor costs." Moreover, California is not alone: similar programs can be found in Utah and Arizona.[64]

The "saving" of labor costs through the use of inmates as firefighters illustrates the functioning of disaster capitalism. It is also an illustration of how the carceral state works hand-in-hand with neoliberalism. The use of inmates is an austerity measure that shifts social and financial responsibility to the most vulnerable in society. Philip Goodman notes advocates for the program see it as a means of rehabilitation and atonement through heroic work. However, he and others have argued exploitation is occurring, with the term "slave labor" being used by activists. While Californians passed Proposition 57, which gave inmates credit for time served, the $2-per-day wage for dangerous work is only part of the dehumanization. Incarcerated firefighters have died on the job, and they have died while training for the job. The questionable ethics and injustice of relying on inmates has raised the concern of civil rights activists and was a concern of prison-strike organizers in 2018.[65]

Given the ever-increasing scale of wildfires, California and other states rely on private firefighting companies as well. In response to the rolling back of the state funding, insurance companies and private contractors have come to fill the gap—for the super-wealthy. As wildfire worsens, it is not surprising that private companies have taken on a greater role in firefighting in the age of neoliberalism. In the United States, the National Wildfire Suppression Association represents over 150 private wildland fire contractors. The organization estimates that 40 percent of firefighting services nationwide come from the private sector.[66] This process began in the 1980s, as the Forest Service was encouraged to downsize seasonal fire crews and rely more on private contractors. This has led to growth in the private sector, which currently consists of companies across the country of varying sizes. Grayback Forestry, for example, fights fires in the manner that one might imagine, deploying fire crews to construct fire lines and extinguish flames. Other companies such as 10 Tanker LLC and Rogers Helicopters provide air support in firefighting activities. These American companies also provide support and services overseas, such as during the 2019 to 2020 Black Summer in Australia.

However, these companies do not simply fight wildland fires. They are increasingly working in a variety of different emergency response areas. GFP Emergency Services and Capstone Fire and Safety, for instance, have firefighters, rescue technicians, and EMTs ready to fight forest fires as well as deal with industrial disasters and other emergencies. The Danish company Falck is one of the largest private "safety" service providers in Europe, offering a wide range of emergency services such as ambulance dispatching and firefighting. Falck not only runs 65 percent of municipal fire brigades in Denmark, it also operates in North America, the Caribbean, and Latin America. In the United States, Falck began operating in Louisiana and Texas in 2008, following their purchase of Alford Safety Systems, which specialized in emergency management services for the oil industry. In 2011, Falck purchased LifeStar Response and Care Ambulance, making it one of the largest ambulance providers in the country. This rise of private companies offering emergency services represents both the expansion of global disaster capitalism and neoliberalism.

This situation is not unlike the metropolitan solid waste situation described in chapter 3. The successful growth in a local market ultimately leads to overseas expansion. Moreover, emergency management services and privatized firefighting involves sophisticated logistical operations centers to coordinate the delivery of people and resources via aircraft or trucks to disaster sites. Unsurprisingly, emergency management and firefighting agencies have logistics officers that assess incidents, order resources, and then mobilize and coordinate efforts.

There are, of course, consequences to the burden being shifted. Public agencies that abdicate their responsibility to protect the public in turn create situations in which human life and labor are devalued. In California, the reliance on inmates and contractors for firefighting allows risk to be shifted to the individual firefighters and their families. Inmates and privately contracted firefighters who are killed often do not have the same death benefits as unionized firefighters working for CalFire.[67] As the goal is to save money and generate profit, we see that labor exploitation also means exploiting taxpayer dollars. Private forestry contractors in Oregon have been accused of using millions of federal dollars to issue H-2B visas to lower labor costs—in fact, forestry contractors have become the largest employers of H-2B visa holders.[68] The top countries of origin for H-2B workers are Mexico, Jamaica, Guatemala, and South Africa. Latinx immigrants (*pineros*) in particular represent a large, invisible workforce facing unsafe working conditions and exploitation.[69] However, this is not just an American phenomenon. Canada has relied on firefighters from Mexico and South Africa to fight its fires. Again, cost savings or exploitation is key, as South African firefighters battling Canada's Fort McMurray Fire were asked to leave following a pay dispute.[70] This exploitative, practically colonial provision of fire suppression and related wildland management services demonstrates firefighting is linked to wealth and power.

Expanding a City's Burning Boundaries

In sum, the physical growth of cities and the consumption associated therewith contributes to wildfire risk. A great deal of urban growth occurs in the wildland–urban interface and other areas where people are directly interacting with brush, foliage, and other flammable material. The increased risk to people and their property is not caused by fire moving toward them, but rather by people moving toward fire. Not only do cities place more people and buildings in harm's way but they fuel the global warming that increases wildfire risk.[71] Cities are sites of production and consumption. Thus, they are major polluters that generate the carbon that warms the planet—through building activities, automobiles, electricity use, and factories.[72] Again, this is a feedback loop that amplifies fire and other hazards.

Urban development and neoliberalism operate in contradiction to effective wildland management. Zoning practices encourage urbanization in the wildland–urban interface, which in turn increases population density, activities that increase fire risk, and climate change–inducing consumption. While the cost of firefighting largely rests on public agencies, neoliberal abdication of responsible provision of public goods has resulted in tax dollars being redirected toward contractors. Put another way, firefighting is not a public good in this approach. Rather, it is reduced to lucrative fire suppression contracts that do not help cities adapt to an increasingly flammable environment.

The result is ever riskier cities. While climate change has increased wildfire risk, this is exacerbated by policymakers allowing ongoing development in hazardous environments and the increased reliance on industries that profit from disaster. The consequence is that people are put at risk, or those who are most vulnerable pay the cost of poor decisionmaking. It is homeowners who are expected to protect themselves by purchasing insurance policies or paying for home renovation, rather than the government providing a public good. Not everyone can pay for protection. As Fraser and Freeman note, the spark that lit the Great Chicago Fire of 1871—O'Leary's Barn—was in a working-class immigrant neighborhood.[73] Fire, like other disasters, affects those who are most vulnerable. It is those at greatest risk who live in the riskiest part of the city.

Effective alleviation of risk requires that society view safety as a public good and not a commodity. It also requires us to move away from simple mitigation of hazards and thinking about how cities need to adapt to a warmer and drier climate and build with resilience and sustainability in mind. However, we also must be vigilant against the co-optation of adaptation, mitigation, resilience, and sustainability. The problem of co-optation, introduced in this chapter, is addressed further in chapter 5.

5
Assessing and Managing Risk
• • • • • • • • • • • • • • • • • • • •

> There is a universal feeling, a universal fear, that our progress in controlling nature may increasingly help to weave the very calamity it is supposed to protect us from.
> —Theodor Adorno, *Negative Dialectics*

Previous chapters have considered how risk is linked to the fiscal decisions of policymakers and corporations as well as the physical construction of the built environment. Certainly, there have been attempts to adapt and to mitigate disaster within the framework of capitalism. However, we must be careful in simply designating this all as greenwashing. Corporations often misrepresent the sustainability of their products. While greenwashing does overlap with the deployment of disaster management plans, it is essential to examine the real practices of financial actors that co-opt adaptation, mitigation, resilience, and sustainability projects. We see that decisionmaking is heavily influenced by the desire to balance risk and economic growth. In turn, we must critically examine how capitalist ideology affects the way cities manage risk.

This chapter aims to do three things. The first is to discuss what capitalists mean by risk reduction. Despite increased risk, neoliberal policymakers and corporations keep building in disaster-prone areas. To deal with risk, municipal institutions increasingly rely on the market: from insurance to financial instruments such as green bonds. This fiscal context informs the way cities shape and manage the built environment. Second, this chapter examines the concepts of adaptation, mitigation, resilience, and sustainability, all of which are commonly used in the study of disaster, risk, and environmental activism in slightly different ways. However, it remains to be seen whether these strategies can overcome the imperative to grow, expand, and profit.

The third issue addressed in this chapter is how risk is turned into a metric and assessed. To make sense of adaptation, mitigation, resilience, and sustainability regarding disaster, I examine existing mechanisms for assessing risk reduction and related environmental management initiatives. Once environmental risks are identified and evaluated, projects are assessed as to whether they can effectively protect people, property, and infrastructure. This may mean using tools established by various organizations, including the International Standards Organization (ISO) and Leadership in Energy and Environmental Design (LEED), that offer "best practices" for companies and government agencies. Indeed, I have been critical of the political-financial structure of capitalism and its ability to deal with urban disaster appropriately. These tools are not—in themselves—necessarily flawed. Rather, the problem is how these tools are used as stand-ins for long-term risk reduction. Such assessment programs transforms actual practices into abstractions disconnected from social reality. In turn, solutions to urban disaster risk leans toward short-term technofixes that do little to tackle the drivers of disaster risk.

The Fiscal Nature of Risk Reduction

In contemporary capitalist cities, the development of the built environment—whether through the private or public sector—involves a great deal of capital or money. Capital improvement projects represent a large portion of municipal budgets before and after a disaster. As such, public finance plays an essential role in shaping the physical nature of cities. Cities raise money for construction projects not only through taxes but also by using a variety of financial tools and instruments, including popular strategies such as tax-increment financing and real estate investment trusts.[1] There are also monetary contracts that are bought and sold among parties. For cities, this typically means bonds for infrastructure projects. More recently, we have seen the development of new financial instruments tied explicitly to nature and disaster, such as catastrophe and green bonds as well as weather futures. The claim is that these instruments help cities raise money and divert risk, thus making them more resilient. As Sarah Knuth explains, use of financial tools have expanded since the Great Recession, as capitalists look for new ways to generate revenue from urban development.[2] Desiree Fields, meanwhile, has argued that financial capital and the urban built environment are interdependent. In fact, neoliberal municipal policymakers are key players in allowing this financialization to take place. Fields gives the example of private equity firms, investment firms, cross-national real estate companies, and other figures in the financial services sector buying up real estate and taking control of public infrastructure.[3] These financial tools ultimately link the physical and fiscal nature of cities to the abstractions of contemporary capitalism. This is the global context in which cities now find themselves. Not only do they live with catastrophe, but urban space is now bound to a financial industry in which everything is commodified.

Acceptance of this commodification is tied to the underlying ideology of endless growth. This governs economic and political decisionmaking under capitalism. It also results in progress being measured in metrics or indicators such as level of urbanization, new homes sold, or gross domestic product (GDP). This adherence to growth is one of the reasons for the popularity of concepts such as decoupling, whereby growth can occur without environmental harm. Intergovernmental organizations such as the Organization for Economic Cooperation and Development (OECD) and the United Nations have included decoupling as a sustainable development goal. The U.N.'s International Resource Panel, for example, has linked decoupling to waste, water, and disaster management programs. The body claims that green growth, decoupled from environmental harm, can reduce exposure to hazards.[4] That said, decoupling has also been subject to criticism. For instance, in 2019, the European Environmental Bureau released a report suggesting that there is no empirical evidence that decoupling or green growth is a realistic strategy for long-term sustainability.[5]

This is because a central premise of decoupling is dematerialization—that is, linking economic growth to less destructive activities. It is reasoned that a shift away from manufacturing to information technology can lead to green growth. However, what we see is the opposite. Technology remains tied to material processes and contributes to urbanization. Wealthy countries depend less on manufacturing because of deindustrialization as related activities are moved overseas.[6] Indeed, wealth generation in the twenty-first century relies on information, finance, and technology. However, all those sectors still require the development of physical infrastructure for communications and an ever-increasing need for energy.[7] At the same time, the development of infrastructure supports the urban growth machine. We see this in the development of peri-urban tech parks, suburban sprawl, and the emergence of so-called smart city programs that fuel gentrification in urban centers. Technology, in other words, has a physical dimension.

The finance services sector, no doubt, plays a significant role in placing cities into riskier and risker environments. This is because insurance and finance are intertwined.[8] The systematic nature of disaster capitalism is visible in the way the financial and insurance industries have commodified risk and worked alongside disaster management firms to maximize the returns on urban growth.⁹ Here, it is essential to consider the nature of insurance. Simply put, insurance is a form of risk management that protects against financial loss. As the U.S. Supreme Court noted in *Helvering v. Le Gierse*, 312 U.S. 531 (1941), risk shifting and risk distribution are central to the insurance industry. This form of insurance is written against actual loss, such as the cost of rebuilding a home. Typically, a policyholder works with brokers or agents who are connected to an insurance company.

The entire insurance sector's operations are about risk management. Insurance companies themselves must redistribute the risk they now possess. Risk shifting means getting someone else to help pay for losses. Behind the scenes, reinsurance helps the original insurance company spread risk by passing on individual policies to a reinsurer. Given that reinsurers help insurers cover claims, reinsurers are providing security to insurance companies, allowing them to offer more policies and collect more premiums that benefit reinsurers. With reinsurance, the original insurance company pays part of the premiums it collects to a reinsurer, which in turn assumes some risk by accepting partial responsibility for losses. While individual insurers might specialize in providing coverage for homeowners, reinsurers create diverse portfolios to balance out different types of risks. Some of these reinsurance companies are well-known investment firms such as Berkshire Hathaway (which in recent years has been ranked as either the fourth or the fifth largest reinsurance company in the world). It also includes well-known insurers such as Lloyd's, based in the United Kingdom. Operating internationally, these companies counterbalance risk by providing coverage for a wide array of areas—such as health, mortgage, and property. This global reach ensures that losses for one catastrophe are offset by premiums coming in from the lack of disaster in another area.¹⁰

American International Group (AIG) is a multinational financial services company with a focus on insurance. In fact, AIG is a major underwriter of commercial insurance policies. AIG's role in the 2008 financial crisis is well documented. Before the Great Recession, the company provided coverage for the debt obligations of other financial institutions involved in mortgage lending. This impacts urbanization, as insurance facilitates the buying and selling of property. Moreover, insurers assist banks in determining who poses a high risk when applying for home loans. For instance, a typical homeowner needs mortgage insurance and disaster coverage. This, in turn, affects who can afford to live in a neighborhood. As Gregory Squires points out, this not only results in redlining and segregation but produces uneven development.¹¹

Insurers also provide coverage for municipal legal liabilities as well as property, and they invest in urban development. According to the National Association of Insurance Commissioners (NAIC), U.S. insurers are also significant investors in municipal bonds, accounting for $3.8 trillion.¹² Because of this relationship between providing coverage and investing in growth, it is unsurprising that companies such as SwissRe were involved in programs such as the Rockefeller Foundation's 100 Resilient Cities initiative. SwissRe not only provided catastrophe modeling for the initiative, it also strongly recommended various risk-transfer solutions and weather insurance products to help cities bounce back from disaster.

Given that insurers are at financial risk if disaster strikes, they increasingly provide consulting services to decisionmakers building and managing cities. In addition to weather insurance, AIG offers services to manage environmental risks such as catastrophe preparedness plans. Other insurance and reinsurance companies provide risk assessments to city-builders that can and do influence how things are built. In fact, American companies such as Travelers and Liberty Mutual offer coverage to cities and other public entities. Insurers, in this way, can exert political and economic pressure on cities and developers to adopt standards and technologies that mitigate disaster damage. The development of building and fire code are an example.[13]

However, following the 2017 wildfire and hurricane seasons in the United States, AIG noted a need to purchase more reinsurance to balance out its risks. Despite insurers' interests in mitigating risk, urban development continues. Political leaders and other businesses are typically more concerned with short-term growth than adapting for the long term.[14] Along these lines, one AIG executive claimed that in the future, the company will "look for opportunities to further reduce volatility in our results, as we position the company for long-term profitable growth."[15] In other words, AIG is looking to mitigate the blow of future catastrophic events while looking for additional revenue.

The development of products that protect cities against hazards ideally allows cities to manage financial risks associated with property damage and economic shutdown. However, Kunreuther and Michel-Kerjan warn that this places our entire economic system at war with the weather, which is now a liability.[16] The goal is to manage weather patterns and the environment in a way not unlike Vegas-style casinos, in which the house rarely loses. In fact, Michael Lewis has called the rise of catastrophe futures "nature's casino."[17] As in a casino, this paradigm necessitates running a variety of different games beyond insurance and reinsurance to cover a potential payout. As discussed in chapter 1, coverage for such policies is supported by various financial instruments. Given the exchange-oriented nature of these instruments, a shift has occurred from ensuring that premiums can cover losses to the idea that the insurer can generate even more profit.

Since the 1990s, we have seen markets trade a greater number and variety of catastrophe bonds, weather derivatives, and futures as part of the global insurance-linked securities market. These financial tools emerged after events such as Hurricane Andrew (1992), the Northridge earthquake (1994), the Chicago heat wave (1995), and the Milwaukee winter (1997–1998) and were developed so that insurance companies could offset the financial risk of a weather extreme or disaster of unprecedented size occurring. The Chicago Mercantile Exchange (CME) began trading weather derivatives at this time, and in the mid-2000s, the CME expanded into hurricanes before starting in 2010 to offer futures on rainfall.[18] Outside the United States, the London International Financial Futures and Options Exchange and Eurex are other exchanges that deal with weather derivatives and futures.

This reliance on the market is distinct from relying on insurance. Insurance or reinsurance only pays out if there is a loss. Instead, the market is made up of exchangeable contracts dependent on weather or disaster events meeting certain conditions before payments are to be made. For example, a company worried about extreme cold could buy a future to mitigate this risk by buying exchange-traded catastrophe products on CME's Heating Degree Day Index. Similarly, if a company is worried about heat, it could buy a product on the Cooling Degree Day Index. Contracts are based on whether weather temperatures are close to the annual averages or exceed expectations. These contractual conditions are not unlike betting on a point spread, typical of oddsmaking for gambling on a sporting event. Oddsmakers calculate the number of points by which a team is expected to win, and gamblers wager on whether the team beats expectations. The seller of a weather derivative uses climate data for a particular city to set prices. The idea is that statistically, temperatures should average out. This is what the seller is betting on. If the actual temperature is stable or normal, the seller profits. However, if it is erratic, then the buyer profits. In this case, buyers are able to insure or hedge against dramatic changes that could affect their business. For buyers, this helps to offset risk if their business is heavily reliant on heating or cooling—such as construction and energy companies, as well as agriculture.

Similarly, catastrophe bonds emerged to address the risk insurers face if a major disaster were to occur. The idea behind a catastrophe bond is that insurance companies have a steady flow of premiums coming in while significant catastrophes are supposed to be relatively rare. Yet, such natural disasters are costly when they occur. If a large-enough catastrophe were to occur, collected premiums might not be sufficient to cover the policies in place. In turn, the insurance company could create a catastrophe bond with the assistance of an investment bank. With catastrophe bonds, if no disaster occurs, the company will pay interest to the investor who purchased the bond. In other words, the investor is betting that cataclysmic events are rare and will make money by the maturity date. If a disaster does happen, then the amount paid by the investor for the bond would be used by the insurance company to help pay claims. In this scenario, the investor would lose money during a disaster, but it would help insurers stay afloat after a significant event.

Increasingly, we see that such instruments provide financial coverage not only for private policyholders but for public agencies that set up disaster funds. The intent is to manage risk and ensure that there is enough money to cover a catastrophe. This includes interlocal risk pools, or the establishment of the disaster relief funds in Florida (discussed in chapter 2). Since the 1990s, Mexico has shifted financial risk to international markets through a mix of reinsurance and catastrophe bonds. Caribbean countries created a multinational insurance pool called the Caribbean Catastrophe Risk Insurance Facility (CCRIF) in 2007. The pool allows multiple countries to pay insurance premiums into a fund that can pay for disaster recovery anywhere in the region. Given that the premise is about risk sharing, the CCRIF has since expanded to Central American countries, helping to broaden the CCRIF's finances. Despite the public nature of the fund, Kevin Grove argues it still reflects financialization of disaster management, as it pools risk with parametric insurance to manage catastrophe.[19] In the same vein, the CCRIF has also begun to rely on reinsurance and catastrophe bonds to distribute risk, as well as on parametric insurance on agriculture and businesses such as fisheries.[20] This model of risk sharing is seen elsewhere. In 2009, the World Bank also created a Multi-Catastrophe (MultiCat) program to help governments—especially in the Global South—create portfolios that pool resources and distribute risk. The program was primarily adapted from Mexico's experience with catastrophe bonds, and it also coordinates with the CCRIF.

Here, it should be noted that insured losses only account for about 25 to 40 percent of total economic loss from natural disasters. Of course, there are human and social costs to disaster as well. This is one of the problems with financialization: It reduces the experience of disaster to economic measures or abstract metrics. According to the Centre for Research on the Epidemiology of Disasters at the Université Catholique de Louvain's database (EM-DAT), while Europe generally experiences greater financial losses from disasters than Africa, a greater number of Africans than Europeans are affected by disaster. Unfortunately, the insurance and reinsurance industry see this as an economic opportunity. SwissRe, notes, "Emerging markets remain the engine of growth for the global insurance market."[21] The global insurance market, like all sectors under capitalism, are obsessed with expansion and the externalization of costs. As such, emerging markets do not just represent more policies and premiums, but they are a way for companies to offset risk.

Insurance, however, is only one means of obtaining the money to deal with a disaster. Emergency loans are another way that money is raised to deal with a disaster. Such loans can come in many forms. For example, there are government programs that provide low-interest loans to homeowners and businesses from declared disasters. In wealthy countries such as the United States, this is facilitated by private banks. The World Bank also offers accessible lines of credit immediately after a disaster, again with the goal of achieving "cost efficiency" for its clients, as well as helping investors find new ways to make money. While loans may be low interest, on a global scale there is a great deal of money to be made. Given this approach, it is unsurprising that major financial firms such as Goldman Sachs, MunichRe, and SwissRe have worked alongside the World Bank Treasury to issue loans to countries that need to manage risk.

Advocates for such programs claim that disaster mitigation is only possible with adequate funding and that both governments and private entities should transfer their risk to the market to minimize budget volatility and grow fund reserves in the case of catastrophe.[22] In addition, cities need to raise funds to support mitigation of future disasters. Many municipal as well as private city-building projects are funded through bonds—essentially loans that raise money for their projects. Green bonds are issued for sustainable development projects—broadly defined. The goal of these bonds is to attract investors to the laudable goal of fighting climate change and building resilient cities. These increasingly popular bonds bring in private capital from around the world to finance energy projects, transportation, waste management, and disaster resilience projects. While cities and other public entities can issue green bonds, multilateral development banks, including the Asian Development Bank (ADB), the European Investment Bank (EIB), and the World Bank, as well as corporations, have issued the majority of green bonds to support development projects.

Given the popularity of green bonds, there has been growing discussion of creating new instruments that facilitate greater environmental, social, and governance (ESG) investment. More recently, there has been a discussion of resilience bonds among financial services companies, insurers, public institutions, and nongovernmental organizations (NGOs). These have a much narrower focus than green bonds in that they target projects that directly deal with natural disaster. Also called municipal or climate adaptation bonds, they support pre-catastrophe development as opposed to post-disaster recovery. Advocates for resilience bonds again include the investment bank Goldman Sachs, the risk management firm RMS, the reinsurer SwissRe, the Rockefeller Foundation, and the design firm Re:Focus. Examples of potential resilience bond projects suggested by Re:Focus include coastal barriers and seawalls in Florida, New Jersey, and Virginia. The firm's projects were chosen based on models generated by RMS of potential economic losses from sea-level changes and how projects in those areas could reduce risk.[23]

The popularity of such bonds coincides with the general trend toward ESG investments. Undeniably, investors see the social and environmental value in this sort of investment. However, there is a practical component as well. It is reasoned that these bonds help manage risk by encouraging investments that counteract climate change, thereby reducing risk. In turn, advocates of resilience bonds note that a lot of support is coming from the insurance sector, which has an interest in reducing risk.[24] For instance, bonds that support seawall construction can allow for new policies to be issued in otherwise precarious places. At the same time, insurers can also invest in bonds as reliable fixed-income investments.

Given the broad areas supported by green bonds, the annual issuance of green bonds currently exceeds $100 billion a year. Green bonds are audited to ensure that they are being used for "green" purposes (see figure 5.1). However, the World Bank concedes that "there is no standardized framework for reporting and measuring the impact of green bonds. This results in challenges regarding the extent to which proceeds are allocated to their intended use or that proceeds have actual positive environmental impact."[25] In response, in 2018, the European Commission's report on financing sustainable growth called for the creation of standards and labels for green financial products.[26]

Despite these concerns, it appears that such strategies for raising money are working—at least in corporate ledgers. We see this in figure 1.2. The amount of money put into bonds, reinsurance, and other financial instruments since 2008 has nearly doubled, no less amid record-setting disasters in recent years.[27] However, the contradictory nature of this growth has not gone without concern. Critical scholars have raised concerns about the development of new financial instruments because of the potential for new problems that they bring. For instance, Dick Bryan and Michael Rafferty criticize financial instruments as "deconstructing the social and economic world into ever more precisely defined attributes, each of which can be configured into a measurable (but sometimes contestable and often fragile) instrument to be priced and traded."[28] In 2013, the European Insurance and Occupational Pensions Authority (EIOPA) noted that "without adequate supervision, such developments could cause systemic risk."[29] In other words, lack of oversight over these instruments has consequences for financial resilience. As such, it is not just regulators who are concerned about regulation. For example, Berkshire Hathaway temporarily pulled back on investing in catastrophe bonds in 2014. Following the devastating 2017 hurricane season, the *Wall Street Journal* suggested that some investors "have been waiting years for a big event to create such a market correction."[30] In turn, some investors were becoming worried that the market for catastrophe bonds and other insurance-linked securities could burst because of all the major disasters that year.

However, we have not seen a step back occur. According to the *Financial Times*, the catastrophe bond market in 2018 grew to $30 billion as investors dug in, despite $136 billion in insured losses worldwide in 2017. Asset management company OppenheimerFunds reported overall growth between 2017 and 2018 and has opted to expand its investment strategies into the catastrophe bond

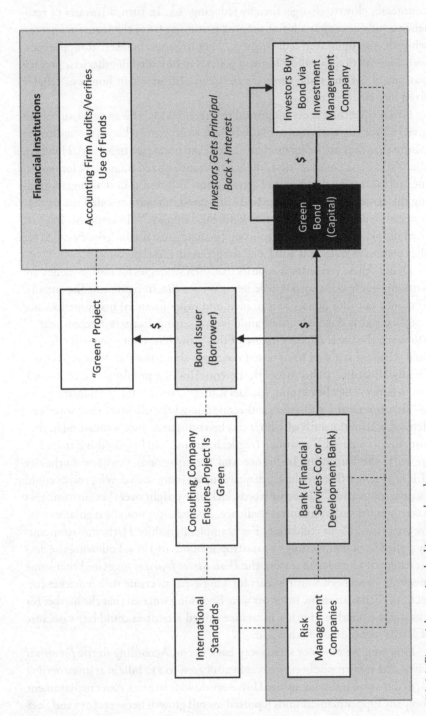

FIG. 5.1 Financial actors involved in green bonds

market.[31] Again, this raises serious concerns as to whether or not a financial bubble could occur as insurers, investment banks, and related industries incorporate catastrophe, green, and resilience bonds in their risk management plans. Much like earlier warnings, the EIOPA noted in 2018 that "a close monitoring of this trend is warranted, however, to examine the risk of a possible green bubble developing."[32] Similarly, the reinsurer SwissRe expressed concern about systemic economic resilience, due to insurance gaps and the structure of the global financial system.[33]

Adaptation and Evolution of Urban Risk

What sort of projects do these bonds support? In recent years, work on natural disasters has focused on adaptation and resilience in regard to protecting lives and property. The term "adaptation" has become more prominent as scholars and practitioners realize the need to cope with climate change, rising sea levels, and increased risk. These approaches are influenced by ongoing developments in the field—for instance, sustainability's popularity has waned in the urban studies literature in favor of resilience.[34] In particular, resilience has become a dominant framework for addressing the growing threats that cities face.[35] While adaptation and resilience refer to long-term approaches to disaster, it appears that the de facto response has been to mitigate harm and call it adaptation or resilience.

Some planners and policymakers argue that adaptation anticipates disaster. Advocates for adaptation suggest that such approaches are also cognizant of the social realities of disaster. Natural disasters and environmental hazards are an outcome of interactions between society and nature. While a site may have particular hazards present, social problems such as inequality and poverty can make the existence and usefulness of various technologies moot. In a world that is warming, real adaptation means city-building with vulnerable populations in mind—children, migrants, women, communities of color, and those in poverty. Technologies that mitigate storm surges, deal with trash, or fireproof homes do little for those who cannot afford them and have no choice but to live in a dangerous environment. This means that communities, cities, and regions must figure out a way to adapt to the current *social* reality of a more volatile environment.

While humanitarianism may influence developments in urban safety, we need to be wary of legitimizing the urban growth machine and the treadmill of production. Take, for instance, sustainability and the drive for green energy sources. Some advocates, such as ecological modernists, believe that we can effectively deal with climate change through technology. As we saw in chapter 3, some industries have used such discourses to "greenwash" metropolitan waste management by claiming that new garbage incineration technologies are sustainable. Rather than adjusting our economic system to value nature's role in urban metabolisms and adapt accordingly, the goal is to develop "greener" ways to increase consumption levels.

Indeed, policymakers are faced with the pressure to support growth and protect private property while minimizing harm to their constituents. As such, cities stuck within the treadmill of production invest resources into mitigation and adaptation strategies that appeal to the financial and real estate sectors while playing lip service to concerns about natural disaster and environmental hazards. The treadmill of real estate takes and develops new and old spaces to generate increasing revenue. As Scott Frickel and James Elliott argue, there is endless churn in cities. Residences and businesses appear and disappear over time, and amid this churn, there is an attempt at risk containment. Agencies will focus on a few readily obvious hazardous sites to reduce public concern. While this gives the impression of effective regulation, other hazardous conditions are ignored. This invisibility is worsened as homes, restaurants, or stores replace these unremedied places.[36]

The idea of churn and risk containment applies not only to the legacy of hazardous sites but to places damaged by disaster. Under contemporary capitalism, mitigation strategies may make buildings, companies, and infrastructure safer and marketable in the short term, but they do not necessarily make the city more resilient to the long-term challenges of changing sea levels, weather patterns, erosion, and other hazardous conditions. In turn, we build and rebuild on risky spaces. It is generally accepted that the application of new technology and management strategies will reduce risk. This is not unlike the environmental Kuznets curve, which assumes that affluence over time leads to a tipping point wherein environmental harm is reduced. In the case of a disaster Kuznets curve, income manifests itself as technology used to mitigate harm. However, the mitigation of harm does not necessarily mean a long-term reduction of risk. This hypothesis has also been put to test in the case of natural disaster by Derek Kellenberg and Ahmed Mobarak, who found that in developing countries, disaster risk increases with growing GDP before later decreasing. This seemingly follows the Kuznets curve. However, Kellenberg and Mobarak note that this generally suggests that less developed countries may have competing interests in hazard mitigation and economic development. Rather, economic development could be creating new risks as countries aggressively modernize. Correspondingly, the curve does not explain local-level mechanisms for creating risk.[37]

Mitigation, Sustainability, and Resilience

Establishing effective disaster plans is undoubtedly complicated, requiring coordination among local, national, and international institutions and stakeholders. At the same time, these actors experience pressure from the urban growth machine's coalition of corporations and policymakers to promote ongoing expansion. In turn, existing risk management strategies have shifted the financial burdens and practical responsibilities of coping with disaster away from banks, developers, and other corporate entities.[38]

While profit drives the financing of such initiatives, this does not mean that no action is taken. It is suggested that greening the built environment in conjunction with careful planning can significantly minimize the harm that cities and city-building cause to the environment. For all parties, there is undoubtedly a need to reduce the immediate consequences—or lessen the blowback—of disaster. Mark Pelling argues that disaster mitigation is often a reactive measure that tends to manifest as engineering fixes that employ physical solutions rather than social ones.[39] Again, technology is not neutral. In an advanced capitalist society, we have a fixation on technology's ability to solve our problems. In earlier chapters, I have criticized mitigation strategies and technology because the deployment of such fixes gives the illusion of safety or sustainability in the near term. Instead, the failure to address the growth-logic of capitalism does little to reduce risk and promote urban resilience in the face of disaster.

No doubt *effectively* implemented mitigation strategies in cities can reduce the impact of disasters. Mitigation, simply put, is a tool, strategy, or technology that minimizes the impact of disasters or reduces risk. This includes the adoption and enforcement of building codes, insurance programs, public awareness campaigns, land-use policies, and hazard mapping. It includes engineering projects, monitoring, communications systems, redesigning infrastructure, and renovating buildings. For instance, "mitigation," according to the Federal Emergency Management Agency (FEMA), involves "analyzing risk, reducing risk, and insuring against risk."[40] All these initiatives sink a great deal of capital into the built environment with the idea that harm is minimized. However, I have previously suggested that mitigation increasingly involves managing risk and hazards as a capitalist enterprise. Alternatively, this could also be understood as protecting economically "productive" elements of the city. Thus, we need to ask whether mitigation projects are short-term technofixes or part of a systemic framework guiding cities toward less risky buildings, infrastructure, and communities.

Addressing resource consumption in the built environment is an example of this approach. Tackling building efficiency is a means of dealing with climate change and natural disaster. Energy consumption related to buildings represents about 76 percent of all electricity use and is responsible for an estimated 45 to 50 percent of carbon emissions. In addition, buildings account for 13.6 percent of potable water consumption and a third of all waste generated around the world.[41] A popular response to building-related environmental harm is the LEED rating and certification program developed by the U.S.

Green Building Council (USGBC) to encourage responsible and sustainable building construction. Hunter Lovins and Boyd Cohen, proponents of "natural" capitalism, point out that green building has become a multibillion-dollar sector that has received great fanfare from Wall Street, real estate developers, major commercial chains, and policymakers. Indeed, such initiatives are increasingly a part of a "climate capitalism" and a growing green economy.[42]

In the case of booming cities, capitalist expansion and green urban development comes in the form of the construction of new energy-efficient homes and shopping centers and the establishment of new special-purpose districts with mitigating designs such as bioretention basins. Some of this development is funded by green bonds or other ESG investments. However, such projects are not necessarily inclusive or accountable forms of urbanism. Kenneth Gould and Tammy Lewis, for example, note that structural mitigation strategies, or engineered solutions, encourage green gentrification. In the absence of sufficient checks on such projects, displacement during construction or increased cost of living can occur. As such, Gould and Lewis found that contributions to environmental sustainability do not necessarily support social sustainability. As capital is sunk into the built environment, the "greening" of existing neighborhoods can force people to move toward even more precarious or hazardous parts of the city.[43]

The legitimization of green city-building is an example of deploying urbanism as planet-saving technology. However, Jesse Goldstein argues that such measures—including a potential Green New Deal—can work to "resuscitate the very social, technological, economic, and political system that is responsible for accelerating (or possibly even causing) our planetary problems."[44] Again, we see that mitigation technologies are often unable to escape the dominance of economic—primarily capitalistic—forces. The development of mitigation systems without consideration of social outcomes renders them hard engineering technofixes rather than truly transformative transitions to a less risky future.

Similarly, Ashley Dawson calls these sort of mitigation projects sinks for surplus capital that create the illusion that speculative real estate is sustainable.[45] Such projects rely on marketable solutions while arguing the ongoing expansion of cities can be safe and green. As discussed previously, proponents of ecological modernization assume that society will adapt to environmental decline and risk by developing new technologies and that consumers will make rational choices that will reduce disaster risk. This could mean more clean and efficient energy to cut greenhouse gas (GHG) emissions, supply-chain management and logistics for garbage and sewage, as well as fire-resistant building materials. This belief system often works together with neoliberalism's emphasis on market solutions to problems. While technological innovation may be disruptive to existing industries, the structure of our social-economic system is not upset. Consequently, such approaches to mitigation and adaptation are particularly ripe for co-optation by disaster capitalism and neoliberal policymaking.

Of course, specific mitigation strategies *can* be useful in hazardous environments. Thus, they have been successfully marketed as part of broader "sustainability" initiatives by policymakers. In 1987, the Brundtland Commission, also known as the United Nations World Commission on Environment and Development, defined sustainable development as "development that meets the needs of the present without compromising the ability of future generations to meet their own needs."[46] Following this approach, sustainability rests on the so-called three Es of economy, environment, and equity. This understanding of economy is fundamentally understood as being in line with capitalism. The commission argues for reorienting technology to mitigate risk and explicitly states innovation can be used to support economic development—especially in developing countries.

Here, it is important to ask what it means to have a clean and safe environment and who has access to it. The Brundtland Commission contends that sustainability initiatives are most effective if there is citizen/stakeholder participation and decisionmaking. In other words, equity is at a very rudimentary level built into the planning and implementation of sustainability projects. Similarly, the United Nations International Strategy for Disaster Reduction (UNIDSR) handbook on resilient cities claims that "sustainability can be achieved if the community itself and local authorities understand the importance of and need for local preparedness and response."[47] Indeed, this and other programs provide actionable items that can encourage stakeholders—for example, residents of a city—to take part in mitigation plans. However, participatory planning practices that involve the entire community do not always occur.

From the practical standpoint of planners, sustainability has meant making the city and policymaking smarter. Jack Ahern has noted the popularity of concepts such as "smart growth" and the call for "new urbanism" in planning in recent decades. These urban design movements have sought to reduce environmental impact by making cities walkable and increasing density. However, Ahern is critical of this notion of sustainability, which he calls a form of "fail-safe" urbanism. This understanding of sustainability, according to Ahern, suggests that city-building *can* be implemented without harming the environment and that we *can* build cities that do not fail—reflecting an anthropocentric hubris. Within Ahern's critique is the question of whether an equilibrium can be reached—or a perfect balance between nature and society.[48] Similarly, Simin Davoudi argue that sustainability is grounded in equilibristic and closed-systems theories and ignores human agency.[49] In other words, a sustainability framework assumes that we can successfully manage environmental problems—as well as decoupling growth and harm.

Again, there are serious doubts as to whether decoupling is realistic. Mindy Benson and Robin Craig suggest that sustainability is no longer a viable goal because of its assumption that natural resources can be effectively managed amid climate change. In other words, the imbalance between nature and society is so great that equilibrium is not possible. Rather, they contend that resilience is a better concept, preparing human society for the coming irreversible changes to our environment.[50] That said, there are different approaches to resilience, including the notion that urban resilience can be engineered through policy and technology. This technocratic approach to urban resilience also emphasizes mitigation and recovery efforts to return a city to its pre-disaster state. As such, Richard Klein and colleagues have questioned the usefulness of resilience in dealing with natural hazards. Returning to a pre-disaster state is problematic because of the wide array of vulnerabilities that existed before and after the disaster.[51] Alternatively, Steward Pickett and colleagues contend that resilience is also associated with the ability to adapt, including coping with internal and external threats. As such, the concept is still useful.[52]

Despite these debates, we have seen the expansion of various disaster- and risk-reduction programs and organizations in recent decades. In turn, definitions matter, as they are often the starting point in establishing metrics and assessments projects. In addition to the private actors mentioned in previous chapters, there are several national and international entities, such as nongovernmental organizations and humanitarian organizations, working on disaster resilience.[53] Given the broad range of actors involved in disaster and risk management, it is not surprising that definitions of resilience and other approaches to dealing with disaster vary. Over the years, the United Nations has endorsed the Yokohama strategy (1994), the Hyogo framework (2005), and the Sendai framework (2015) to reduce disaster losses around the world. In each framework, the U.N. has not only laid out an agenda for making people safer but also redefined what it means to be resilient. The Yokohama strategy focused on "prevention, mitigation, preparedness and relief." Since then, Hyogo has emphasized disaster resilience, and Sendai has focused on risk reduction. Sendai has moved away from the framework of "disaster management" to a focus on risk as a societal responsibility rather than just a governance problem.

Correspondingly, various arms of the United Nations, such as its Development Programme (UNDP) and International Strategy for Disaster Reduction (UNISDR), have all encouraged the development of resilient cities in a variety of ways. For these bodies, the core of resilience is "the ability of a system, community or society exposed to hazards to resist, absorb, accommodate, adapt to, transform and recover from the effects of a hazard in a timely and efficient manner, including through the preservation and restoration of its essential basic structures and functions through risk management."[54] They have encouraged large-scale solutions such as reducing carbon emissions as well as addressing local and regional risks to water supplies, agriculture, and food security and limiting development in places that have known disaster vulnerabilities. Importantly, there is also an emphasis on assessing and understanding the prevalence of the different risks that contribute to a disaster.

These definitions matter in that the U.N. plays a role in humanitarian and disaster relief assistance. There are networks of nongovernmental and civil society organizations that operate between supranational bodies such as the United Nations and the local level. For example, the London-based Global Network of Civil Society Organizations for Disaster Reduction (GNDR) seeks to "increase community resilience and reduce disaster risk around the world."[55] There is also the Community Practitioners' Platform (CPP) that helps grassroots groups working on local-level resilience projects connect to one another, with a particular focus on empowering women. These groups often work in conjunction with the UNIDSR, especially in the Global South, to connect different groups to resources, know-how, and each other

In addition to the U.N., humanitarian organizations such as the International Red Cross and Red Crescent, known for their post-disaster relief, are now also encouraging pre-disaster resilience projects. The Red Cross and Red Crescent define resilience as "the ability of systems (and people) to effectively respond and adapt to changing circumstances and to develop skills, capacities, behaviors and actions to deal with adversity—'resilience' can be described as a process of adaptation before, during and after an adverse event." Accordingly, the organizations have begun "developing and improving its resilience approaches and programming by combining the humanitarian concern for imminent threats with the sustainable and longer-term approaches and institutional strengthening traditionally associated with development."[56]

Finally, there are also groups such as the Global Facility for Disaster Reduction and Recovery (GFDRR), which, alongside the World Bank, has a "City Resilience Program" to assist local governments in finding financial resources with which to facilitate investment in disaster resilience, which GFDRR defines as "a community's ability to respond to, cope with, and recover from a disaster." Specifically, the City Resilience Program "is an effort to consolidate city-level engagement across key sectors by providing a consistent range of resilience-building investment options in areas of infrastructure, systems and finance."[57] However, it should be noted that the program relies a great deal on capital investment planning. In its 2018 annual report, GFDRR emphasized its mobilization of "private capital experts support cities to develop a pipeline of resilience-building investments and expand the financing options available to them."[58]

Amid these debates over the idea of resilience, what is perhaps most important is what social actors do with the concept. Lawrence Vale has encouraged us to critically examine what resilience means for different actors working on disasters. Vale notes that management analysts, engineers, and ecologists use the term in different ways.[59] Indeed, social actors in this area—including activists, architects, planners, scientists, and of course, communities—have embraced approaches such as biomimicry, biophilic cities, and cradle-to-cradle.[60] Such plans represent a different way of thinking about urban nature and therefore perhaps create possibilities for making resilient cities. Yet, from a critical standpoint, we need to step back and examine how these actors operate within the context of expansion-oriented capitalism. Consequently, Leitner and colleagues argue that actors are often producing commodified norms that reflect a "neoliberal governance agenda in resilience clothing."[61]

Defining Resilience through Environmental Management and Assessment

David Beer contends that a key component of neoliberalism is the expansion of metrics and assessment in all areas of society, which not only allows a means of measuring productivity but also reinforces claims regarding the effectiveness of economic strategies and risk management.[62] In fact, metrics and assessment have become powerful governing agents in environmental management. The GFDRR encourages a move "towards risk assessments that can guide decision-makers towards a resilient future."[63] Indeed, we have seen the expansion of different models for assessing risk, one of which involves establishing frameworks, metrics, and standards that allow the assessment of whether a project is resilient. The USGBC, for instance, has been working to establish standards for urban resilience much like its work on LEED. In 2017, the USGBC announced that it would be deploying the RELi resilience standard and rating system developed by the Institute for Market Transformation to Sustainability (MTS) and Capital Markets Partnership (CMP) as a certification system for the planning and design of buildings and communities.

In addition to the USGBC standards, there are several assessment tools that are designed to assist in the mitigation of environmental harm and risk. For example, the Building Research Establishment Environmental Assessment Method (BREEAM) in the United Kingdom and Green Star in Australia have established rubrics for assessing a building's contribution to environmental sustainability as well as fire safety and security. Notably, these standards do consider the well-being of building users, residents, and neighbors. Along these lines, there is also the Eco-Management and Audit Scheme (EMAS) for the European Union, which is adapted from the International Standards Organization (ISO) 14000 family of standards. EMAS and ISO 14000 are designed to help organizations design and implement environmental management programs. ISO 14000/31000 guidelines also offer advice and means of assessing how human stakeholders fit into environmental impact and risk plans.[64]

While these standards are designed to be applicable in several settings, they are most commonly considered in construction and building operations. Construction is the fourth largest sector gaining ISO 14000 certification.[65] Construction, after all, has a direct impact on the environment. It has been suggested that ISO and other systems formalize environmental and risk-management practices across the industry and embed sustainable practices within organizations. Key to the standard is the establishment of environmental management systems that allow organizations to reach their own goals.[66] Yet the organizational focus of environmental assessment tends to shift the focus from human stakeholders to entities such as businesses.

While such standards, in theory, consider the importance of unique local social and economic realities, it is debatable how much this occurs, because these standards construct uniformity across time and space. This is because supranational organizations, transnational reinsurance firms, banks, and NGOs, mentioned previously, have made assessment and metrics ubiquitous in city-building, urban governance, and risk management. Stefan Timmermans and Steven Epstein point out that the creation of international standards is representative of a "modernizing and globalizing world."[67] Indeed, standards for environmental and risk management are developed by experts. Yet, as Scott Knowles reminds us, experts and their work exist within a social and historical context. Today, this context is one of ever-expanding disaster capitalism. We cannot disassociate the rise of metrics and assessment from the context of neoliberal governance.[68]

Ultimately, we need to examine the actual practice of city-building and ask whether such measures make cities safer: Do these metrics measure the effectiveness of mitigation, sustainability, and resilience projects? Safety standards do overlap with building codes and other means of creating safer cities; however, they should not be mistaken for ideology-free city-building and risk management. A building built to code in a floodplain is still at risk. Consequently, the standards and related policy decisions do not necessarily challenge the status quo. Rather, they often cooperate with or reinforce existing understandings of city-building. Resilience plans often contend that urban sustainability and risk reduction initiatives can offer synergistic fixes for policymaking pressures such as economic imperatives, interurban competition, and regulation.[69]

We see this in the UNIDSR's handbook on creating disaster-resilient cities. It notes the benefits of investing in disaster risk reduction and resilience, citing economic growth and job creation. It also mentions a potential for an "increased tax base, business opportunities, economic growth and employment as safer, better-governed cities attract more investment"[70] Therefore, it is not uncommon for policymakers to advocate synergistic "green" public–private partnerships when discussing urban development plans, such as brownfield redevelopment, coastal real estate projects, or tourist destinations.[71]

The outcome is that corporations can use standards to claim that they employ "green" or "resilient" practices in a competitive marketplace. Corporations increasingly use ESG principles to claim a positive social and environmental impact as well as to encourage long-term growth in investments. Recently, the ISO has noted that its standards are the "secret to unlocking green finance."[72] It has claimed that its standards provide a transparent and clear means of defining what it means to be green, especially as new financial instruments, such as green bonds, that require auditing emerge. A reliance on market solutions shifts environmental governance to private sector and nongovernmental organizations that are not accountable to the public.

The use of standards in this fashion—not unlike scientific advisory committees—is often neither transparent nor democratic. Again, while experts in a variety of areas may offer useful and valuable insight into the drafting of such standards, they are not necessarily the ones using them. Given the significance of corporations in the modern world, we cannot help but be concerned with the selective adoption of green and risk-mitigation strategies. In fact, the neoliberal acceptance of standards such as ISO, LEED, and other assessment tools may undermine attempts by the state to establish enforceable regulations. As in other areas, we are seeing that nonstate, market-driven compliance systems have gained enormous authority in recent decades despite opaque top-down decisionmaking.[73]

Another reason that such standards do not necessarily create safer or more sustainable cities is that they do not address systemic drivers of urban disaster *risk*. The RELi standard encourages resilience, restoration, regeneration, sustainability, and wellness. The goal is shock-resistant buildings and communities that are able to repair and replenish themselves to thrive and maintain their future potential. Once deployed, RELi, like LEED, will have a point system that allows organizations and agencies to assess the resiliency of buildings and communities. They help the designers of specific projects and real estate companies by assisting in the development of sustainable and resilient cities.[74]

Is overall risk is reduced? Here, we can consider LEED as a similar case. A study by the National Research Council Canada found that while LEED buildings used less energy per floor than their counterparts, their overall energy consumption was higher.[75] In other words, LEED and other programs increase efficiency but do not necessarily decrease consumption. This is a prime example of the Jevons paradox, the model by which technology improves resource efficiency yet increases consumption because of increasing demand.[76] Similarly, in the case of disaster mitigation, the technology improves the safety of specific buildings and structures, but it can also facilitate risky urban growth. This results in a built environment that is endlessly relying on new mitigation technologies because the original problems of expansion and climate change were never addressed.

Conclusion

Regenerative Urbanism

> In the end it has made physical disintegration—through war, fire, or economic corrosion and blight—the only way of opening the city up to the fresh demands of life.
> —Lewis Mumford, *The City in History*

While disaster risk has the attention of policymakers, we see that that disaster alone is not enough for them to abandon expansion-oriented urbanization. By 2050, the world's urban population will be about double what it is now. This means that two-thirds of the world's people will be living in urban areas. An estimated 90 percent of this growth will occur in Africa, Asia, Latin America, and the Caribbean. Notably, global megacities will increase their populations and expand their social, political, and economic reach. Not only do we see this contributing to climate change, but it is placing people into increasingly risky environments.

Historically, people have built cities alongside bodies of waters for agriculture, trade, and drinking. Now we see rising sea levels and floods. In turn, urban populations—especially the poor—are placed on the frontline of disaster. Among the most severe risks under contemporary capitalism are those generated by climate change. Global warming, certainly, is a driver of urban risk. Climate change leads to catastrophic storms and sea levels rising. It means hotter urban heat islands and flammable hinterlands. Global warming has dire consequences for human well-being, especially for the poor, and affects the basic functioning of cities. As such, we need to transition to a low-carbon economy. The decarbonization of the world economy, at a basic level, means reducing emissions by moving away from fossil fuels, accompanied by de-growth. Unfortunately, existing strategies fall short of the necessary benchmarks recognized by the 2016 Paris Agreement and affirmed by climate scientists. The reality is that the political economy of the modern world is deeply entrenched in fossil fuels.[1]

No doubt a reorientation toward less risky urbanism requires a shift from fossil fuels. However, that needs to occur alongside a restructuring of our economic system. As I have argued throughout this book, capitalism will always create new forms of fiscal and physical risk. The problem is capitalism's obsession with expansion and how it produces an ideology in which long-term benefits are brushed aside for short-term profit. A resilient or regenerative project will always be under-resourced within this system. Finance and related services such as private banking and insurance will always direct resources toward profit-seeking, thus supporting endless urban growth. Corporate environmental, social, and governance (ESG) principles will not get us to less risky cities. Correspondingly, a Green New Deal will see limited success if it cannot resist or operate outside of private finance. However, we should not lose hope.

I have presented a critical look at how cities have dealt with risk and disaster. *Risky Cities* has looked at how the built environment, disaster, and capitalism shape modern cities. Previous chapters laid out a critique of disaster capitalism and how environmental *bads* become exchangeable *goods*. While new technologies may mitigate damage, capitalism hijacks these tools. Neoliberal fiscal policies have allowed corporations to co-opt resilience projects and undermined attempts at cleaner and more equitable city-building. The spatial fixity of cities and the need for capital to aggressively seek profit drives the political legitimization of risky city-building. Neoliberal faith in the marketplace ignores a history of risk simply being redistributed, while social and environmental protections are undermined.

In other words, we have not changed course. As such, we need to examine whether specific programs or strategies reflect a proactive regenerative approach to urbanism, as opposed to a reactive means of dealing with environmental risk and disaster.[2] Indeed, the development of building codes, metropolitan solid waste management, fire-resistant building materials, and advancements in engineering have reduced the loss of life and property. However, the building of homes and construction of infrastructure that allow for growth does not necessarily mean that the foundations of cities are stable. Previously, I illustrated how profit-seeking motives, rather than benevolence, drive such strategies. New means of handling waste have simply increased consumption. Developments in fire safety are used to legitimize building homes in flammable topographies.

The result is larger cities that are sinking, increasing their waste generation, or burning.

The presence of hazards is bound to the historical development of cities in physical space. For instance, the development of seaports and industries such as fishing and tourism coincide with city-building along a coast. The fact that buildings and infrastructure exist along ever-changing coastlines means that there will be risks. Human settlements are built on vulnerable sites, and our interventions increase risks such as erosion, flooding, and exposure to storms. In turn, cities have engineered solutions to protect human settlements, such as seawalls that dramatically alter the natural and built environment. While engineering solutions reduce harm, they do not solve the problems of ongoing coastal development that serve real estate and financial interests. Instead, such solutions produce a false sense of security while creating more risk by placing more people and buildings in harm's way. At the same time, blowback from bad policy reveals the contradictions of disaster capitalism. For instance, there currently is greater awareness of social vulnerabilities because of the humanitarian crises we have seen post-disaster. This can provide the opening for us to change the trajectory of urbanization.

Contradictions can create opportunities. Contradictions include profit-seeking activities that increasingly disconnect the city from nature, creating new forms of risk or the potential for disaster. Capital adapts by discovering ways to profit from the exploitation of nature—specifically by commodifying disasters. This results in another contradiction. Disaster capitalism is reliant on contradictory processes that undermine urban nature; in turn, cities become more fragile. Today, we see this in the ongoing destruction of nature and how it increasingly reveals environmental injustices. Capital may have figured out how to benefit in the short term by commodifying environmental *bads*, but the ongoing destruction of our ecosystem will ultimately result in a dead end for the disaster-industrial complex. It is precisely these contradictions that create openings that allow for resistance against the commodifying forces of disaster capitalism. It is here that the built environment and nature can be reclaimed from capitalism, and we can establish urban natures that more effectively deal with natural hazards.

Indeed, anti-sprawl and smart-growth or no-growth initiatives have emerged amid growing concern for the environment.[3] In the United States, there have been calls for a Green New Deal, whereby the national economy would be redirected toward environmentally and socially conscious projects.[4] Some of these initiatives directly challenge the dominant ideology of growth and expansion in capitalist society, largely due to growing awareness of our current climate crisis. Yet, given the power of financial capitalism, this resistance must—by necessity—be radical rather than reformist. For instance, as corporations have adopted so-called ESG investment philosophies, a Green New Deal could face co-optation like many of the strategies and technologies discussed in earlier chapters. Indeed, various initiatives are marketed as resilient and can withstand the onslaught of a natural disaster. Yet this market focus, as we have seen, only redistributes risk. Risk becomes a tradable commodity, and it becomes a way for financial capital to generate more profit. As a result, those who are already most vulnerable find themselves experiencing disasters at their worst—while corporations make money off infrastructure breakdowns, exposure to waste or wildfire, or building collapses from earthquakes and sinkholes.

In this conclusion, I would like to advance theory in search of solutions and provide hope for a future with less risky cities. We must radically rethink how public institutions and the built environment can serve real city dwellers. While financial institutions influence how we make cities, they can be challenged and we can create alternative ways to fund the building of less risky cities. Ongoing urban struggles and social movements can inspire change, of course. In a world in which cities are sinking and the planet is literally burning, it has become obvious that the capitalism fuels disaster. Such injustices prompt resistance and create the potential for change. We see pushback in cities, whereby public services, spaces, and nature are focal points. Rather than accept the risks of capitalism, people are fighting back.[5]

Indeed, urban space and its production are not neutral; rather, urbanization under modern capitalism is a tool in the process of accumulation and expansion. The abstraction of space leads to the commodification of environmental *bads*. In the words of Lefebvre, this results in architects and urbanists caught in a "world of commodities."[6] Checkmarks, environmental assessment plans, representations such as algorithms, and short-term solutions replace safety. Resilience, as I argued in chapter 5, is turned into a checklist for construction companies and risk managers. The result is urban space that operates as a tool and product of disaster capitalism. A stepping-stone against the co-optation of resilience is taking back urban space. Therefore, urban space, and the right to the city, must be recovered so that inhabitants may be able to live in less risky cities.

This change of trajectory requires a reframing of city-building through the lenses of regenerative urbanism whereby concepts such as "cultivation" and "nurturing" replace expansion-oriented growth. Here, I would argue that a regenerative form of urbanism focuses on differential needs. Other scholars have proposed regenerative urbanism as a project that is equally urban and environmental that encourages self-renewing systems.[7] A regenerative city would not be free of hazards or risk. However, regenerative cities could be structured to take into consideration a variety of ecological, geographic, and social rhythms from both the past and present.

Time, for starters, is an important consideration for regenerative urbanism. We can only break the pattern of capitalist appropriation by learning from disaster. This requires memory of the past and planning for the future. Christof Mauch notes that disaster—while devastating—is not remembered in the same way as other horrific events such as war.[8] Earthquakes and hurricanes leave a mark, but when new events occur, they are forgotten as the damage is fixed, plants grow back, or new crises overtake the news cycle. As such, solutions deal with singular disaster "events" as opposed to the more complicated phenomena of urban development and growth. The building of less risky cities should be not just a reaction but a proactive response to problems that we know existed, continue to exist, and are likely to worsen if we do not change course. From rebuilding to the erection of monuments, such practices can regenerate post-disaster cities while making remembrance a part of the social and mental landscape.[9]

Looking to the future, after all, requires learning from the past. In Greek drama, the *catastrophe* is the resolution to the story. It often includes the protagonist being driven to change by a process of self-discovery and overcoming hubris or arrogance. The catastrophe, however, does not necessarily mean a happy ending. For cities, disaster does drive *change*. It leads to rebuilding and transformations of the built environment. Whether cities *learn* to cope with catastrophe effectively is a more complicated story. Recovery, according to Kathleen Tierney and Anthony Oliver-Smith, is affected by "macro-societal and even global factors."[10] While they are speaking of material and social recovery efforts, such macro-level forces also influence what lessons are (or are not) learned. Indeed, cities may want to adapt and mitigate risk. Yet capitalism has appropriated and assimilated risk-reducing tactics to advance its obsession with growing and expanding cities, thus undercutting effective adaptation to climate change and disaster. In other words, we are not learning.

Taking back urban space and reducing risk requires a holistic understanding of the city. Cities change. However, can they effectively adapt? In many ways, they have. Yet the lessons learned from risk and disaster are often partial. Politicians, corporations, and many city dwellers still see economic growth as a desirable goal. Economic development, it is argued, leads to disaster-resistant technologies that solve our problems. However, this is often without regard for the potential blowback from such strategies. Because of this myopia, market failures are often ignored. Decisionmakers overlook asymmetrical growth and inequality while simultaneously creating new threats.

This means we need to fix past mistakes and engage in regenerative urbanism. David Harvey, drawing on Lefebvre, argues that cities have a tendency toward "restoration."[11] Indeed, cities have adapted to waste problems and disaster events by—at times—building atop old landfills and rubble. There are instances where we have done much more than just reclaim the space. There are cases of public space and nature being restored. For instance, Fresh Kills Landfill in Staten Island, New York, is now Freshkills Park. César Chávez Park in Berkeley, California, and Sai Tso Wan Recreation Ground in Hong Kong, as well as Chambers Gully Reserve in Adelaide, Australia, were previously waste disposal sites. They are examples of regenerative urbanism in that they address existing hazards and rejuvenate urban space.

Yet not all projects that reclaim damaged urban space are built for the public. For instance, investors have begun seeing brownfields as an investment opportunity given their prevalence in cities where land is expensive. By developing real estate in previously contaminated areas, they can open otherwise closed-off areas of the city for development. For instance, in 2017, it was announced that 1,680 housing units would be built atop a former landfill in Santa Clara County in California.[12] Given that housing units represent urban growth, it is questionable as to whether this project can represent a form of regenerative urbanism.

Here we see that the rejuvenation of urban nature can go either way. Similarly, William Freudenburg highlights the difference between disasters in producing corrosive and therapeutic communities.[13] Disaster capitalism and the inequality that it produces have a corrosive dimension. At the same time, disaster events can play a role in collective memory and positively reshape the built environment as well as notions of community. Crises, after all, represent moments in which communities and societies may perpetuate the status quo or change trajectories. Correspondingly, the aftermath of a crisis can lead to therapeutic outcomes such as restoring public spaces.

Coming up with solutions is undoubtedly a complicated endeavor. The core of my critique is the growth-logic of capitalism. More specific than "growth," it is expansion that creates a precarious urban situation. The rebuilding of cities and a reliance solely on mitigation technologies can make us safer in the short term. However, the deployment of such strategies does not tackle the ideological and fiscal drivers of increased precarity. Risk is not merely tied to growing populations. Rather, it is expanding human settlements and related economic activities that exacerbate risk and environmental harm. It is the physical and fiscal nature of disaster capitalism. As such, we must reconfigure our understanding of urban nature by removing expansion-oriented economic activities from the day-to-day functioning of cities.

In addition to learning form the past, what should regenerative urbanism look like? What does this mean for cities and disaster? Again, the work of Henri Lefebvre is useful in trying to understand how this is possible. Lefebvre offers the alternative of "differential space," which reopens everyday life to new possibilities, as a counterpoint to capitalist space.[14] For Lefebvre, differential space is not so much a fixed site as much as a product of ongoing political practice and resistance against the homogenizing influence of capitalism. Michael Leary-Owhin has argued extensively that differential space should have a place in planning theory. Differential space offers us a framework for understanding how the city can resist the growth-logic of capitalism.[15] This means changing our social and economic system so that urban life can occur without creating new environmental risks.

In other words, regenerative urbanism should consist of multi-scalar and multi-temporal approaches to fight the homogenizing effects of global capitalism. Part of this would be the recognition and reclamation of past uses of space.[16] Doing so would also allow for the cultivation of both society and nature within urban space. Applying differential space to regenerative urbanism can be thought of as a ground-up process. It would look at how city dwellers have used urban space in the past and the present. Unlike top-down notions of decoupling economic growth from environmental harm, heterogenous social uses of space are recognized. In this way, differential space is like biodiversity. It is grounded in understanding variety and variability within a system.

When we talk about recognizing the multi-scalar and multi-temporal dimensions of cities, there needs to be consideration of the before, during, and after of disaster across the different spaces. We also must consider continuity and persistence in the built environment. Uta Hassler and Niklaus Kohler argue that within the built environment, different time scales operate for different people and places. This is owing to differential access to natural, physical, and social capital. City dwellers may have differential access to the city, and neighborhoods vary in their history and structures.[17] As such, regenerative urbanism requires the building of cities, communities, and disaster plans that recognize different people and spaces and that they cope with situations differently. This is because environmental hazards affect people and communities differently.

Thus, we need to think about the both short- and long-term timescales of disaster vulnerability. For instance, Kate Aronoff and colleagues noted that a diversity of governance and economic structures would be integral to a successful housing plan in a Green New Deal. For example, neighborhood housing can serve as a local resiliency center during a disaster.[18] Key here is that communities are empowered and given resources to address their unique social and environmental problems. This means building up a community's social infrastructure and capital. Daniel Aldrich, for instance, argues that social capital is integral to resilience. He has argued that social capital can operate as nonfinancial insurance that provides immediate assistance and the potential for rebuilding post-disaster.[19] For this to work, urban institutions must nurture existing social relationships and cultivate new forms of social capital. Similarly, Eric Klinenberg contends that social infrastructure is critical not just during a disaster but before. Such investments are not just about catastrophe; they are also about environmental and public health.[20]

This means that the social and environmental dimensions of city-building come first. For instance, Christopher Béné and colleagues remind us of the relationship between poverty and vulnerability. They argue that it may be useful to decouple measures of resilience from specific environmental hazards and link it directly to vulnerability.[21] Vulnerability, as stated earlier, exists regardless of the risk. As such, the criteria for effective, or strong, decoupling at the urban level shifts the focus of disaster-related engineering projects of building defensible spaces to emphasizing communities—or real people and neighborhoods—that are threatened. By placing people first, the precautionary principle would play a more significant role in the way in which natural hazard risk is approached. In particular, there would be an awareness of the diverse range of stakeholders and the various vulnerabilities that people face.

A vulnerability focus, in turn, lends itself to social solutions rather than just hard engineering solutions. In practice, this could take the form of participatory urban planning, whereby the stakeholders are involved in reshaping the city. This would emphasize programs that nurture the local built environment as opposed to technofixes from technocrats. This may also mean the advancement of solutions that may not be traditionally marketable, such as conservation, cultural preservation, more inclusive cities, and counteracting anthropogenic climate change.[22]

This change would also manifest itself in people reclaiming the urban built environment, as well as nature, allowing a diverse range of city dwellers their "right to the city." Lefebvre, for instance, has suggested that urban space can be "diverted, reappropriated, and put to a purpose quite different from its initial use."[23] For Lefebvre, this would make the city a living place full of possibilities. In this alternative future, human life would be cultivated. Put another way, regenerative cities will not be singular defensible spaces serving disaster capitalism. Instead, they will be multifaceted spaces that deal with problems of vulnerability by nurturing physically and socially strong communities.

An issue is whether "growth" can be decoupled from its capitalistic understanding—whereby the increased rate of accumulation or profit margins is the sole measure growth.[24] The shift to an alternative economic ideology and system is imperative, as it is the only real way to move away from growth-oriented urban development. There is no doubt that the expansion of urban land through construction is a core component of urbanization. Building cities alters the natural environment and regional hydrology, causes deforestation, and draws on natural resources. This is easier said than done. Cities play an essential role in economic growth in Africa, Asia, and Latin America. There is, of course, the ethical dilemma of denying "development" and "growth" to the Global South, while the North has reaped the benefits of environmental exploitation for centuries.

However, growth and investment can be replaced by the value-orientations of "cultivation" and "nurturing." This is different from arguing for de-growth. Instead, resources should be devoted to economically, environmentally, and socially beneficial projects such as human and planetary health, as well as the fight against inequality. Such an orientation would mean not only developing an urban infrastructure that supports both nature and society but making that relationship core to how we govern, plan cities, and deal with disaster. As such, by moving away from expansion-oriented urbanism, we can perhaps create global cities that are more just and less risky.

It has been argued that because of constrained budgets, the risks governments cannot manage should be shifted elsewhere—for example, to reinsurance, regional risk books, and catastrophe bonds. This financial bind is arguably a manufactured problem directly linked to the fiscal decisions of neoliberal politicians. As such, the tendency of capitalism to financialize every aspect of society, as well as the business-friendly concessions offered to corporations when developing resilient cities, must be resisted. Simply put, disaster and risk management should not be left in the hands of corporations. Leaving the door open to this corporate influence is like the Intergovernmental Panel on Climate Change (IPCC) leaving the door open to geoengineering. It allows profitability to get in the way of real solutions.[25] It also allows companies to use ESG to greenwash what they are doing or deflect from the negative consequences of their actions. As David Harvey points out, "environmental and social reasons" often mask land grabs driven by financial interests.[26] In turn we need to reconceptualize our social and economic systems. This means pushing for more democratic and equitable urban governance. For example, the development of participatory forms of budgeting is a starting point in accomplishing this goal.

In a perfect world, an alternative system of finance would be that of a solidarity economy. That is, we would have a system in which quality of life is emphasized for those within society. The focus of institutions working to solve problems—for example, natural disaster—would be on social vulnerability not just financial risk. In such a system, credit unions and public banks, as opposed to corporate banks, would be significant players in providing funds for climate adaptation, disaster resilience projects, and regenerative urbanism. It has also been suggested that public finance is a pathway to an effective Green New Deal. For instance, state investment banks can provide long-term financing to green projects in an otherwise risky marketplace. The idea is that a publicly owned banking system that is democratically accountable would encourage the development of projects that are obligated to reduce disaster vulnerability and provide effective post-disaster relief and rebuilding for stakeholders.

Public financing solutions can also include state-run insurance funds. Public insurance pools would prioritize everyone, rather than just shareholders that demand a high return on investment.[27] The plans offered would not be last-resort catastrophe insurance. By providing comprehensive national coverage, government insurance pools would distribute risk while providing coverage to more people. In addition, this would mean there would be incentives to have meaningful land-use policy as well as effective adaptation and resilience plans. Similarly, participatory budgeting can reshape how public institutions handle disaster preparedness and recovery. Rather than rely on massive insurance companies to handle post-disaster recovery, stakeholders are directly involved in allocating much-needed resources to the areas that are hardest hit.

Technology would undoubtedly still be a part of building safer communities. However, it would be driven by a different starting point—one in which people and nature come before capital. For this to occur effectively, there needs to be a move away from proprietary data, in which information is housed in opaque black boxes only accessible to those with financial resources, and toward open-source data and tools. This would allow for greater democratization of the information necessary for cities, as well as citizens, to address problems of vulnerability.

From the standpoint of planning, reconfigurations of urban nature away from exploitation to balanced and regenerative urbanism could include biophilic cities and cradle-to-cradle practices.[28] In the case of the wildland–urban interface, there have been calls to think more critically about community and other social forces with respect to land use and resource management.[29] For example, rethinking urban nature and using plants (e.g., mangrove trees) would allow coastlines to adapt to climate change; plants would also serve as carbon sinks. This, of course, includes limiting development and establishing urban forests. Such approaches represent a different way of thinking about urban nature and making less risky and regenerative cities.

In conclusion, regenerative urbanism involves more than reducing harm to nature. It also involves building less risky cities. Regenerative urbanism brings into focus a city's environmental impact beyond its formal boundaries. Projects would be interurban, regional, and global and would cultivate and nurture urban nature to reduce risk.[30] This does not simply mean creating a safe habitat for human beings; it also involves creating built environments that enhance or restore the existing "nature." In recognizing the interconnected nature of cities, the goal would be to enrich the lives of city dwellers as well as ensure that planetary health is not put at risk. This is possible, and collectively, we can lay the foundation of a less risky future. As such, I believe we can live in safer, cleaner, and greener cities than currently exist.

Acknowledgments

Risky Cities represents the culmination of my works on cities since graduate school. In turn, this book was made possible by countless people that I have worked with over this period of time. I apologize in advance to those not explicitly named here. First, I am indebted to my mentor Martin Murray and to professors Benita Roth, Richard Lee, Çağlar Keyder, Tom McDonough, Nancy Um, Abidin Kusno, and Anthony King for their guidance during my time at Binghamton University. I also want to thank my friend Utku Balaban for our collaborations and his insights over the years.

Next, I want to express gratitude to my colleagues and students from Boğaziçi University and the University of Gondor; they shaped my global approach to cities. In addition, many other colleagues, conference session attendees, editors, and peer reviewers have shaped my work over the years. Specifically, I wish to acknowledge Andrew Berzanskis, Matthew Clement, Riley Dunlap, Sherry Gerstein, Tim Haney, Vincent Nordhaus, Sean Murphey, and Eric Tagliacozzo for their input on this project from proposal to its current form. In addition, I thank my editor Peter Mickulas at Rutgers University Press for his enthusiastic support of this project.

A sabbatical leave from Kutztown University made this book possible. In turn, I want to acknowledge my colleagues who served on the sabbatical leave committee. As a faculty member at a regional-comprehensive university, I especially want to thank those colleagues who value academic research and were a source of encouragement while I was teaching eight or more courses per year. Specifically, I thank Kurt Friehauf, Mauricia John, and Khori Newlander for their valuable comments on this book's early drafts, as well as Jason Crockett, Bill Donner, and Joleen Greenwood for their support over the years.

I must also recognize my family. No doubt, growing up in California shaped my view of natural disasters. I want to express special appreciation to my parents Jack and Dawn Fu, and my brother Stanley. I especially thank my wife, Sema Hande Öğütcü-Fu, for her patience and love throughout my academic career. Finally, I dedicate this book to my daughter Lara. There is no doubt we live in risky times. I hope that she comes to live in a safer, cleaner, and greener world than the one she was brought into.

Notes

Introduction

1. Aon Benfield data, cited in Matthew Lerner, "Insured Catastrophe Losses Top $20 Billion in First Half: Report," *Business Insurance*, July 19, 2017, http://www.businessinsurance.com/article/20170719/NEWS06/912314577/Insured-catastrophe-losses-22-billion-dollars-Aon-Benfield-Impact-Forecasting; GFDRR, *Investing in Urban Resilience: The Report* (Washington, DC: Global Facility for Disaster Reduction and Recovery, 2015), https://www.gfdrr.org/en/investing-urban-resilience-protecting-and-promoting-development-changing-world.
2. Lerner, "Insured Catastrophe Losses Top $20 Billion in First Half"; GFDRR, *Investing in Urban Resilience*.
3. Fritz-Julius Grafe and Harald A. Mieg, "Connecting Financialization and Urbanization: The Changing Financial Ecology of Urban Infrastructure in the UK," *Regional Studies, Regional Science* 6, no. 1 (2019): 496–511.
4. Henri Lefebvre, *Introduction to Modernity* (London: Verso, 1995), 144.
5. Aon Benfield, "Reinsurance Market Outlook: Reinsurance Proves Its Worth," Aon plc, January 2018, http://thoughtleadership.aonbenfield.com/Documents/20180103-ab-analytics-rmo-january.pdf.
6. Harvey Molotch, "The City as a Growth Machine: Toward a Political Economy of Place," *American Journal of Sociology* 82, no. 2 (1976): 309–332; Harvey Molotch and John Logan, "Tensions in the Growth Machine: Overcoming Resistance to Value-Free Development," *Social Problems* 31, no. 5 (1984): 483–499; Allan Schnaiberg, *The Environment: From Surplus to Scarcity* (New York: Oxford University Press, 1980).
7. Klein's "shock doctrine" includes everything from natural disaster to political upheaval. Indeed capitalism, or more specifically neoliberalism, is a component. However, we also need to look at the specificity of physical threats. See Naomi Klein, *The Shock Doctrine: The Rise of Disaster Capitalism* (New York: Picador, 2010), 176; also see Naomi Klein, *This Changes Everything: Capitalism vs. the Climate* (New York: Simon and Schuster, 2014).
8. Antony Loewenstein, *Disaster Capitalism* (New York: Verso Books, 2015); Kevin F. Gotham and Miriam Greenberg, *Crisis Cities: Disaster and*

Redevelopment in New York and New Orleans (New York: Oxford University Press, 2014); John C. Mutter, *The Disaster Profiteers: How Natural Disasters Make the Rich Richer and the Poor Even Poorer* (New York: Macmillan, 2015).

9   Ulrich Beck, *Risk Society: Towards a New Modernity* (London: SAGE, 1992), 21; also see Ulrich Beck, "From Industrial Society to the Risk Society: Questions of Survival, Social Structure and Ecological Enlightenment," *Theory, Culture & Society* 9, no. 1 (1992): 97–123.

10 Ilan Kelman, "Lost for Words amongst Disaster Risk Science Vocabulary?," *International Journal of Disaster Risk Science* 9 (2018): 6; also see Benjamin Wisner, Piers Blaikie, Terry Cannon, and Ian Davis, *At Risk: Natural Hazards, People's Vulnerability and Disasters* (Oxford: Routledge, 1994).

11 See International Standards Organization, *Risk Management—Principles and Guidelines, ISO 31000:2009*, 1–2, and also *Quality Management Systems—Requirements, ISO 9001:2015*.

12 ISO, *Risk Management—Principles and Guidelines, ISO 31000: 2018*.

13 Amy A. Quark, "Scientized Politics and Global Governance in the Cotton Trade: Evaluating Divergent Theories of Scientization," *Review of International Political Economy* 19, no. 5 (2012): 895–917; also see Amy A. Quark, *Global Rivalries: Standards Wars and the Transnational Cotton Trade* (Chicago: University of Chicago Press, 2013).

14 Lefebvre, foreseeing this, notes that "one of the results of man's control over nature is that he begins to be aware of the aleatory. But this awareness does not leave him powerless and resigned. He integrates the aleatory so that it becomes part of his consciousness and his actions. The new is always surprising, but as long as its newness is not intolerably new, it becomes part of established structures, balances and self-regulating processes"; see Lefebvre, *Introduction to Modernity*, 204.

15 Martin J. Murray, *The Urbanism of Exception: The Dynamics of Global City Building in the Twenty-First Century* (Cambridge: Cambridge University Press, 2017), 9.

16 Mark Pelling, *The Vulnerability of Cities: Natural Disasters and Social Resilience* (London: Earthscan, 2003), 15; Timothy Recuber, *Consuming Catastrophe: Mass Culture in America's Decade of Disaster* (Philadelphia: Temple University Press, 2016); Alissa Cordner and Eliana Schwartz, "Covering Wildfires: Media Emphasis and Silence after the Carlton and Okanogan Complex Wildfires," *Society & Natural Resources* 32, no. 5 (2019): 489–507.

17 The EPA's Disaster Debris Recovery Tool includes over 6,000 facilities. The vast majority are private recyclers and landfills. See EPA, "Planning for Natural Disaster Debris," EPA 530-F-19-003 (Washington, DC: Office of Resource Conservation and Recovery, April 2019), 18, https://www.epa.gov/sites/production/files/2019-05/documents/final_pndd_guidance_0.pdf.

18 John Murawski, "Waste from Hurricane Florence Continues to Overwhelm Landfills," *The News & Observer*, February 2019.

19 Chelsea Shannon, "Officials Decide Where Camp Fire Debris Will Go," *ABC 10*, January 24, 2019, https://www.abc10.com/article/news/officials-decide-where-camp-fire-debris-will-go/103-5df96611-b559-44a8-9e2a-949c5f4f871e; May Lanski, "Illegal Debris Removal in Paradise," *KHSL News*, April 26, 2019, https://www.actionnewsnow.com/content/news/Illegal-debris-removal-in-Paradise-509132501.html; Anne Makovec, "Kincade Fire Clean Up Efforts

Continue with Hazardous Waste Sweep," *CBS SFBayArea*, November 12, 2019, https://sanfrancisco.cbslocal.com/2019/11/12/kincade-fire-clean-up-efforts-continue-with-hazardous-waste-sweep/; Tyler Silvy, "Sonoma County Tells Landowners That Contractors Must Handle Fire Debris Removal," *The Press Democrat*, November 12, 2019, https://www.pressdemocrat.com/article/news/sonoma-county-tells-landowners-that-contractors-must-handle-fire-debris-rem/.

20 EPA, "Planning for Natural Disaster Debris."
21 CBS News, "6 Months after Deadliest Wildfire in California, Water Found with Elevated Levels of Cancer-Causing Benzene," *CBS News*, May 16, 2019, https://www.cbsnews.com/news/camp-fire-california-paradise-unusable-water-with-elevated-levels-of-cancer-causing-benzene/.
22 Serhan Cevik and Guohua Huang, *Fiscal Policy: How to Manage the Fiscal Costs of Natural Disasters* (Washington, DC: International Monetary Fund, 2018).
23 Sarah L. Quinn, *American Bonds: How Credit Markets Shaped a Nation* (Princeton, N.J.: Princeton University Press, 2019), 1.
24 Ken-Hou Lin and Megan Tobias Neely, *Divested: Inequality in Financialized America* (New York: Oxford University Press, 2020), 1.
25 John Foster, "The Financialization of Accumulation," *Monthly Review* 62, no. 5 (2010): 14–15.
26 Manuel B. Aalbers, "Financial Geography III: The Financialization of the City," *Progress in Human Geography* 44, no. 3 (2020): 595–607; also see Robert W. Lake, "The Financialization of Urban Policy in the Age of Obama," *Journal of Urban Affairs* 37, no. 1 (2015): 75–78.
27 Michael Bloomberg, Anne Hidalgo, and Eduardo Paes, "We've Been Mayors of New York, Paris and Rio. We Know Climate Action Starts with Cities," *The Guardian*, May 5, 2016, https://www.theguardian.com/commentisfree/2016/may/05/mayors-new-york-paris-rio-climate-action-cities.
28 United Nations, "Global Investors Mobilize Action in Wake of Paris Agreement," *United Nations Sustainable Development Goals* (blog), January 27, 2016, https://www.un.org/sustainabledevelopment/blog/2016/01/global-investors-mobilize-action-in-wake-of-paris-climate-agreement/.
29 United Nations, "Paris Agreement: Sustainable Development Goals Knowledge Platform," accessed January 2, 2019, https://sustainabledevelopment.un.org/frameworks/parisagreement.
30 United Nations, "Global Investors Mobilize Action."
31 Schnaiberg, *The Environment*.
32 Ideas in chapter 1 were previously introduced in an article in *Environmental Sociology*. The discussion of logistics found in chapter 3 was first explored in an article published in *Urban Studies*. Chapter 4 includes work on wildfires that was covered in pieces published in *Critical Sociology* and *Environment and Planning A*.
33 William Cronon, *Uncommon Ground: Rethinking the Human Place in Nature* (New York: W. W. Norton, 1996), 50. In addition, my work on wildfire was influenced by Bad Religion, "Los Angeles Is Burning," track 6 on *The Empire Strikes First*, Epitaph Records, 2004.
34 This operationalization of theory is influenced by Neil Fligstein and Doug McAdam, "Toward a General Theory of Strategic Action Fields," *Sociological Theory* 29, no. 1 (2011): 1–26.
35 There is a difference between sociological studies of the built environment and the discipline's traditional focus on people. See Matthias Gross, "Human Geography

and Ecological Sociology: The Unfolding of a Human Ecology, 1890 to 1930—and Beyond," *Social Science History* 28, no. 4 (2004): 575–605; Anthony D. King, *Buildings and Society: Essays on the Social Development of the Built Environment* (London: Routledge & Paul Kegan, 1980); John Logan and Harvey Molotch, *Urban Fortunes: The Political Economy of Place* (Berkeley: University of California Press, 1987).

36 Katharine Haynes, John Handmer, John McAneney, Amalie Tibbits, and Lucinda Coates, "Australian Bushfire Fatalities 1900–2008: Exploring Trends in Relation to the 'Prepare, Stay and Defend or Leave Early' Policy," *Environmental Science & Policy* 13, no. 3 (2010): 185–194.

37 J. Lee Jenkins, Edbert B. Hsu, Lauren M. Sauer, Yu-Hsiang Hsieh, and Thomas D. Kirsch, "Prevalence of Unmet Health Care Needs and Description of Health Care–Seeking Behavior among Displaced People after the 2007 California Wildfires," *Disaster Medicine and Public Health Preparedness* 3, no. S1 (2009): S24–S28.

38 Kathleen Tierney, "Disaster Governance: Social, Political, and Economic Dimensions," *Annual Review of Environment and Resources* 37, no. 1 (2012): 341–363.

39 Dianne de Guzman, "Sinkholes, Landslide Warnings in Santa Rosa amid Upcoming Storms," *San Francisco Chronicle*, November 13, 2017, https://www.sfgate.com/bayarea/article/sinkhole-santa-rosa-fires-melted-pipes-rain-12351785.php.

40 Waqas H. Butt, "Distributing Destruction," *City* 21, no. 5 (2017): 614–621.

41 G. Robbert Biesbroek, Rob J. Swart, and Wim G. M. van der Knaap, "The Mitigation–Adaptation Dichotomy and the Role of Spatial Planning," *Habitat International* 33, no. 3 (2009): 230–237.

Chapter 1   Living with Disaster and Capitalism

1 Martin Coward, "Against Anthropocentrism: The Destruction of the Built Environment as a Distinct Form of Political Violence," *Review of International Studies* 32, no. 3 (2006): 419–437; Martin Coward, *Urbicide: The Politics of Urban Destruction* (New York: Routledge, 2008); Benjamin Wisner, Piers Blaikie, Terry Cannon, and Ian Davis, *At Risk: Natural Hazards, People's Vulnerability and Disasters* (Oxford: Routledge, 1994).

2 John Bellamy Foster, "Marx's Theory of Metabolic Rift: Classical Foundations for Environmental Sociology," *American Journal of Sociology* 105, no. 2 (1999): 366–405; John Bellamy Foster, Brett Clark, and Richard York, *The Ecological Rift: Capitalism's War on the Earth* (New York: NYU Press, 2011); Jason W. Moore, "Environmental Crises and the Metabolic Rift in World-Historical Perspective," *Organization & Environment* 13, no. 2 (2000): 123–157.

3 Kevin F. Gotham, "Re-Anchoring Capital in Disaster-Devastated Spaces: Financialisation and the Gulf Opportunity (GO) Zone Programme," *Urban Studies* 53, no. 7 (2016): 1362–1383.

4 Thomas I. Palley, "Financialization: What It Is and Why It Matters," in *Financialization* (Springer, 2013), 17–40.

5 Henri Lefebvre, *Rhythmanalysis: Space, Time, and Everyday Life*, ed. Gerald Moore (London: Continuum, 2004), 8.

6 Henri Lefebvre, *Everyday Life in the Modern World*, trans. Sacha Rabinovitch (New York: Continuum, 2002), 145.

7   Henri Lefebvre, *The Production of Space* (Oxford: Basil, 1991), 26.
8   David Demeritt, "What Is the 'Social Construction of Nature'? A Typology and Sympathetic Critique," *Progress in Human Geography* 26, no. 6 (2002): 767–790.
9   Richard Sennett, *The Conscience of the Eye: The Design and Social Life of Cities* (New York: Alfred A. Knopf, 1990), 41–70; also see William Cronon, "The Trouble with Wilderness: Or, Getting Back to the Wrong Nature," *Environmental History* 1, no. 1 (1996): 7–28.
10 Erik Swyngedouw, "The City as a Hybrid: On Nature, Society and Cyborg Urbanization," *Capitalism Nature Socialism* 7, no. 2 (1996): 65–80; Matthew Gandy, "Cyborg Urbanization: Complexity and Monstrosity in the Contemporary City," *International Journal of Urban and Regional Research* 29, no. 1 (2005): 26–49.
11 Louis Wirth, "Urbanism as a Way of Life," *American Journal of Sociology* 44, no. 1 (1938): 2–3.
12 Riley E. Dunlap and William R. Catton, "Struggling with Human Exemptionalism: The Rise, Decline and Revitalization of Environmental Sociology," *The American Sociologist* 25, no. 1 (1994): 5–30; Nikolas C. Heynen, Maria Kaika, and Erik Swyngedouw, eds., *In the Nature of Cities: Urban Political Ecology and the Politics of Urban Metabolism* (New York: Routledge, 2006); Paul Robbins, "Obstacles to a First World Political Ecology? Looking Near without Looking Up," *Environment and Planning A* 34, no. 8 (2002): 1509–1513.
13 Homer Hoyt, *One Hundred Years of Land Values in Chicago: The Relationship of the Growth of Chicago to the Rise of Its Land Values, 1830–1933* (Washington, DC: Beard Books, 2000).
14 Anne Whiston Spirn, *The Granite Garden: Urban Nature and Human Design* (New York: Basic Books, 1984).
15 Fernand Braudel, *Civilization and Capitalism, 15th–18th Century: The Wheels of Commerce* (Berkeley: University of California Press, 1982); Matthew Gandy, "Landscapes of Disaster: Water, Modernity, and Urban Fragmentation in Mumbai," *Environment and Planning A* 40, no. 1 (2008): 108.
16 Reyner Banham, *Los Angeles: The Architecture of Four Ecologies* (Berkeley: University of California Press, 1971).
17 James C. Fraser, "The Relevance of Human Geography for Studying Urban Disasters," *Space and Culture* 9, no. 1 (2006): 14–19.
18 "Fire in the Home Is Easier to Preempt than to Extinguish," *Fire and Water Engineering*, 1922.
19 Adam Diamond, "Treadmill Acceleration and Deceleration: Conflicting Dynamics within the Organic Milk Commodity Chain," *Organization & Environment* 26, no. 3 (2013): 298–317; Daniel J. Zarin, "Searching for Pennies in Piles of Trash: Municipal Refuse Utilization in the United States, 1870–1930," *Environmental Review* 11, no. 3 (1987): 207–222; Edward Humes, *Garbology: Our Dirty Love Affair with Trash* (New York: Avery Trade, 2013).
20 George Judson, "What Were They Thinking? When Yesterday's Clean Fill Is Today's Toxic Waste," *New York Times*, May 22, 1993, NY/Region, https://www.nytimes.com/1993/05/22/nyregion/what-were-they-thinking-when-yesterday-s-clean-fill-is-today-s-toxic-waste.html; Janet Raloff, "Dirty Little Secret: Asbestos Laces Many Residential Soils," *Science News* 170, no. 2 (2009): 26–28.
21 LaTasha Givens and Julie Wolfe, "New Legislation Could Put Teeth in Trash Pit Sinkhole Accountability," *WXIA-TV/11Alive*, May 4, 2016, http://www.11alive

.com/article/news/local/holding-powerful-accountable/new-legislation-could-put-teeth-in-trash-pit-sinkhole-accountability/170655998.
22. Ulrich Beck, "From Industrial Society to the Risk Society: Questions of Survival, Social Structure and Ecological Enlightenment," *Theory, Culture & Society* 9, no. 1 (1992): 97–123; Dennis S. Mileti, *Disasters by Design: A Reassessment of Natural Hazards in the United States* (Washington, DC: John Henry Press, 1999); Kathleen J. Tierney, "Toward a Critical Sociology of Risk," *Sociological Forum* 14 (1999): 215–242.
23. Naomi Klein, *The Shock Doctrine: The Rise of Disaster Capitalism* (New York: Picador, 2010), 14, also 537.
24. Bruce G. Carruthers and Jeong-Chul Kim, "The Sociology of Finance," *Annual Review of Sociology* 37, no. 1 (2011): 239–259.
25. Kate Booth, "Profiteering from Disaster: Why Planners Need to Be Paying More Attention to Insurance," *Planning Practice & Research* 33, no. 2 (2018): 211–227.
26. Karl Marx, *Capital: Volume I* (London: Penguin, 1990), 929.
27. For an excellent review of the Frankfurt School's relevance in environmental sociology, see Ryan Gunderson, "Environmental Sociology and the Frankfurt School 2: Ideology, Techno-Science, Reconciliation," *Environmental Sociology* 2, no. 1 (2016): 64–76.
28. Henri Lefebvre, *Introduction to Modernity* (London: Verso, 1995), 125; also see Derek Sayer, *The Violence of Abstraction: The Analytic Foundations of Historical Materialism* (Oxford: Basil Blackwell, 1987); Christian Marazzi, *Capital and Language: From the New Economy to the War Economy* (Los Angeles: Semiotext(e), 2008).
29. Cathy O'Neil, *Weapons of Math Destruction: How Big Data Increases Inequality and Threatens Democracy* (New York: Crown, 2017).
30. Sheila Jasanoff, *The Fifth Branch: Science Advisers as Policymakers* (Cambridge: Harvard University Press, 1994); also see Alf Hornborg, *The Power of the Machine: Global Inequalities of Economy, Technology, and Environment* (Walnut Creek, CA: Rowman Altamira, 2001).
31. Sean Mooney, "Reinsurers Should Beware Cat Model Overreliance: Allianz Re," *Global Reinsurance*, November 6, 2014, https://www.globalreinsurance.com/reinsurers-should-beware-cat-model-overreliance-allianz-re/1410647.article; "The Future of Cat Modelling," *Global Reinsurance*, March 3, 2013.
32. David Harvey, *The Urbanization of Capital: Studies in the History and Theory of Capitalist Urbanization* (Baltimore: John Hopkins University Press, 1985); David Harvey, *Spaces of Hope* (Berkeley: University of California Press, 2000).
33. Friedrich Engels, *Dialectics of Nature* (London: Wellred, 2012); Neil Smith, *Uneven Development: Nature, Capital, and the Production of Space* (Athens: University of Georgia Press, 2008).
34. Lewis Mumford, *The Pentagon of Power* (New York: Harcourt Brace Jovanovich, 1974), 11.
35. Albert S. Fu, "Global Wildfire and Urban Development Blowback from Disaster Capitalism," in *Systemic Crises of Global Climate Change: Intersections of Race, Class and Gender*, ed. Phoebe Godfrey and Denise Torres (New York: Routledge, 2016), 225–237.
36. Friedrich Engels, "The Part Played by Labor in the Transition from Ape to Man," in *German Socialist Philosophy*, ed. Wolfgang Schirmacher (New York: Continuum, 1997), 234; also see Hongbo Yang, Thomas Dietz, Wu Yang, and Jindong

Zhang, "Changes in Human Well-Being and Rural Livelihoods under Natural Disasters," *Ecological Economics* 151 (2018): 184–194.

37   JoAnn Carmin, Kathleen Tierney, Eric Chu, Lori M. Hunter, J. Timmons Roberts, and Linda Shi, "Adaptation to Climate Change," in *Climate Change and Society: Sociological Perspectives*, ed. Riley E. Dunlap and Robert J. Brulle (New York: Oxford University Press, 2015), 164–198.

38   Faranak Miraftab, "Neoliberalism and Casualization of Public Sector Services: The Case of Waste Collection Services in Cape Town, South Africa," *International Journal of Urban and Regional Research* 28, no. 4 (2004): 874–892; Erik Swyngedouw, "The Mirage of the Sustainable 'Smart City': Planetary Urbanization and the Spectre of Combined and Uneven Apocalypse," in *Cities in the 21st Century*, ed. Oriol Nel-lo and Renata Mele (New York: Routledge, 2016), 134–143.

39   Adrian Parr, *Hijacking Sustainability* (Boston: MIT Press, 2009).

40   James R. O'Connor, *Natural Causes: Essays in Ecological Marxism* (New York: Guilford Press, 1998).

41   Foster, "Marx's Theory of Metabolic Rift."

42   Jason W. Moore, *Capitalism in the Web of Life: Ecology and the Accumulation of Capital* (New York: Verso Books, 2015).

43   Klein, *The Shock Doctrine*; Antony Loewenstein, *Disaster Capitalism* (New York: Verso Books, 2015); John C. Mutter, *The Disaster Profiteers: How Natural Disasters Make the Rich Richer and the Poor Even Poorer* (New York: Macmillan, 2015).

44   Leigh Johnson, "Catastrophic Fixes: Cyclical Devaluation and Accumulation through Climate Change Impacts," *Environment and Planning A* 47, no. 12 (2015): 2503–2521.

45   Karl Marx, *Capital: Volume II* (Moscow: Progress Publishers, 1956), 104, https://www.marxists.org/archive/marx/works/1885-c2/index.htm.

46   Lewis Mumford, *The City in History* (San Diego, CA: Harcourt, 1961), 444.

47   Richard Hornbeck and Daniel Keniston, "Creative Destruction: Barriers to Urban Growth and the Great Boston Fire of 1872," *American Economic Review* 107, no. 6 (2017): 1365–1398. As a side note, some traditional societies will use prescribed burns to remove old or dead overgrowth to encourage new plants to grow. This practice will be discussed further in chapter 4 in the context of tensions with modern brush management practices.

48   M. Gottdiener, *The Social Production of Urban Space*, 2nd ed. (Austin: University of Texas Press, 1994); David Harvey, "The Right to the City," *New Left Review* 53 (September/October 2008): 23–40; Don Mitchell, *The Right to the City: Social Justice and the Fight for Public Space* (New York: Guilford Press, 2003).

49   Kevin F. Gotham and Miriam Greenberg, *Crisis Cities: Disaster and Redevelopment in New York and New Orleans* (New York: Oxford University Press, 2014).

50   U.S. Department of State, "Status of Post-Earthquake Recovery and Development Efforts in Haiti" (Washington, DC: Bureau of Public Affairs, December 31, 2014), https://www.state.gov/reports/status-of-post-earthquake-recovery-and-development-efforts-in-haiti/; Deborah Sontag, "Earthquake Relief Where Haiti Wasn't Broken," *New York Times*, July 5, 2012, http://www.nytimes.com/2012/07/06/world/americas/earthquake-relief-where-haiti-wasnt-broken.html; Mutter, *The Disaster Profiteers*.

51   Scott Knowles, *The Disaster Experts: Mastering Risk in Modern America* (Philadelphia: University of Pennsylvania Press, 2012), 8.

52. Mark Skidmore and Hideki Toya, "Do Natural Disasters Promote Long-Run Growth?," *Economic Inquiry* 40, no. 4 (2002): 664–687; Hideki Toya and Mark Skidmore, "Economic Development and the Impacts of Natural Disasters," *Economics Letters* 94, no. 1 (2007): 20–25.
53. James R. Elliott and Matthew Clement, "Natural Hazards and Local Development: The Successive Nature of Landscape Transformation in the United States," *Social Forces* (2017): 851–876.
54. Kevin F. Gotham, "Creating Liquidity out of Spatial Fixity: The Secondary Circuit of Capital and the Subprime Mortgage Crisis," *International Journal of Urban and Regional Research* 33, no. 2 (2009): 355–371.
55. Gotham, "Re-Anchoring Capital in Disaster-Devastated Spaces."
56. David Changnon and Stanley A. Changnon, "Major Growth in Some Business-Related Uses of Climate Information," *Journal of Applied Meteorology and Climatology* 49, no. 3 (2010): 325–331; Patrick L. Brockett, Mulong Wang, and Chuanhou Yang, "Weather Derivatives and Weather Risk Management," *Risk Management and Insurance Review* 8, no. 1 (Spring 2005): 127–140.
57. Mark Duffield, *Development, Security and Unending War* (Cambridge: Polity, 2007); Anoja Wickramasinghe, "Tsunami: Building the Nation through Reciprocity While Reconstructing the Affected Areas in Sri Lanka," *Local Environment* 10, no. 5 (2005): 543–549.
58. Raytheon, "Raytheon: Storm School for Everyone—Raytheon Offers Training Modules to Build a Weather-Ready Nation," *Raytheon.com*, accessed December 22, 2018, https://www.raytheon.com/news/feature/storm-school-everyone; also see Naomi Klein, *This Changes Everything: Capitalism vs. the Climate* (New York: Simon and Schuster, 2014), 9.
59. Sarah Scoles, "How Smallsats Could Make a Big Difference for NASA and NOAA," *Wired*, accessed January 24, 2018, https://www.wired.com/story/how-smallsats-could-make-a-big-difference-for-federal-science/; "Global Small Satellite Market Industry Analysis and Forecast 2018–2023," accessed August 5, 2018, https://www.upmarketresearch.com/reports/Global-Small-Satellite-Industry-Market-Analysis-Forecast-2018-2023.
60. Jennifer Alsever, "Space Startups Are Booming," *Fortune*, February 20, 2017, http://fortune.com/2017/02/20/space-startups-travel-satellites/.
61. Leigh Johnson, "Catastrophe Bonds and Financial Risk: Securing Capital and Rule through Contingency," *Geoforum* 45 (2013): 30–40; also see Patricia Grossi, Howard Kunreuther, and Don Windeler, "An Introduction to Catastrophe Models and Insurance," in *Catastrophe Modeling: A New Approach to Managing Risk*, ed. Patricia Grossi and Howard Kunreuther (New York: Springer Science & Business Media, 2005), 24.
62. Joe Palca, "Betting on Artificial Intelligence to Guide Earthquake Response," *NPR.org*, April 20, 2018, https://www.npr.org/2018/04/20/595564470/betting-on-artificial-intelligence-to-guide-earthquake-response; Toh Ting Wei, "How This Startup Is Breaking New Ground in Earthquake Insurance," *Tech in Asia*, October 13, 2016, https://www.techinasia.com/talk/startup-breaking-ground-earthquake-insurance.
63. Susan Strange, *Casino Capitalism* (Manchester: Manchester University Press, 2015).
64. Scott C. Sheridan and Laurence S. Kalkstein, "Progress in Heat Watch–Warning System Technology," *Bulletin of the American Meteorological Society* 85, no. 12

(2004): 1931–1941; Heleen L. P. Mees, Peter P. J. Driessen, and Hens A. C. Runhaar, "'Cool' Governance of a 'Hot' Climate Issue: Public and Private Responsibilities for the Protection of Vulnerable Citizens against Extreme Heat," *Regional Environmental Change* 15, no. 6 (2015): 1065–1079.

65 NYC Emergency Management, *NYC's Risk Landscape: A Guide to Hazard Mitigation* (New York: City of New York, November 2014).

66 There is also the Hazus toolkit—originally produced by the Federal Emergency Management Agency (FEMA)—that has been adapted for open-source projects.

67 "A Smarter Jab," *The Economist*, October 14, 2010, http://www.economist.com/node/17258858; Bourree Lam, "Vaccines Are Profitable, So What?," *The Atlantic*, February 10, 2015, https://www.theatlantic.com/business/archive/2015/02/vaccines-are-profitable-so-what/385214/.

68 Chalmers Johnson, *Blowback: The Costs and Consequences of American Empire* (New York: Macmillan, 2001), 17.

69 Andrew T. Price-Smith, *Contagion and Chaos: Disease, Ecology, and National Security in the Era of Globalization* (Cambridge, MA: MIT Press, 2008), 2.

70 Sam Fankhauser and Raluca Soare, "Strategic Adaptation to Climate Change in Europe," EIB Working Papers No. 2012/01, Luxembourg, 2012, http://www.econstor.eu/handle/10419/88098.

71 Chester W. Hartman and Gregory D. Squires, eds., *There Is No Such Thing as a Natural Disaster: Race, Class, and Hurricane Katrina* (New York: Routledge, 2006); Eric Klinenberg, *Heat Wave: A Social Autopsy of Disaster in Chicago* (Chicago: University of Chicago Press, 2003).

72 Abraham Benavides and Sudha Arlikatti, "The Role of the Spanish-Language Media in Disaster Warning Dissemination," *Journal of Spanish Language Media* 3 (2010): 41.

73 Matt Rahn, "Initial Attack Effectiveness: Wildfire Staffing Study," Wildfire Research Report, San Diego State University, Summer 2010, accessed December 12, 2014, https://www.iaff.org/10News/PDFs/cdffirereport.pdf; Konane M. Martinez, Anna Hoff, and Arecela Nunez-Alvarez, "Coming Out of the Dark: Emergency Preparedness Plan for Farmworker Communities in San Diego County," National Latino Research Center, California State University San Marcos, January 2009, https://www.csusm.edu/nlrc/documents/report_archives/fw-emergency-plan-5-17-10.pdf.

74 Andrew Kirby and A. Karen Lynch, "A Ghost in the Growth Machine: The Aftermath of Rapid Population Growth in Houston," *Urban Studies* 24, no. 6 (1987): 587–596; Mohamed Hamza and Roger Zetter, "Structural Adjustment, Urban Systems, and Disaster Vulnerability in Developing Countries," *Cities* 15, no. 4 (1998): 291–299.

75 Peter S. Alagona, "What Makes a Disaster 'Natural'?," *Space and Culture* 9, no. 1 (2006): 77–79; David L. Brunsma and J. Steven Picou, "Disasters in the Twenty-First Century: Modern Destruction and Future Instruction," *Social Forces* 87, no. 2 (2008): 983–991; Klinenberg, *Heat Wave*.

76 Alice Fothergill and Lori A. Peek, "Poverty and Disasters in the United States: A Review of Recent Sociological Findings," *Natural Hazards* 32, no. 1 (May 2004): 89–110.

77 M. Mirza, "Climate Change and Extreme Weather Events: Can Developing Countries Adapt?," *Climate Policy* 3, no. 3 (September 2003): 233–248.

78 Rolf Lidskog and Daniel Sjödin, "Extreme Events and Climate Change: The Post-Disaster Dynamics of Forest Fires and Forest Storms in Sweden," *Scandinavian Journal of Forest Research* 31, no. 2 (2016): 148–155.

79 Walter Firey, "Sentiment and Symbolism as Ecological Variables," *American Sociological Review* 10, no. 2 (1945): 140.

Chapter 2 Sinkholes and the Risky Foundations of Cities

1 Ashley Dawson, *Extreme Cities: The Peril and Promise of Urban Life in the Age of Climate Change* (London: Verso Books, 2017), 38.
2 Timothy Recuber, *Consuming Catastrophe: Mass Culture in America's Decade of Disaster* (Philadelphia: Temple University Press, 2016), 19, also 36.
3 Mario Parise, "A Procedure for Evaluating the Susceptibility to Natural and Anthropogenic Sinkholes," *Georisk: Assessment and Management of Risk for Engineered Systems and Geohazards* 9, no. 4 (2015): 272–285.
4 Phillip R. Kemmerly, "Sinkhole Hazards and Risk Assessment in a Planning Context," *Journal of the American Planning Association* 59, no. 2 (1993): 221.
5 William R. Catton and Riley E. Dunlap, "Environmental Sociology: A New Paradigm," *American Sociologist* 13, no. 1 (1978): 41–49; Riley E. Dunlap and William R. Catton, "Struggling with Human Exemptionalism: The Rise, Decline and Revitalization of Environmental Sociology," *American Sociologist* 25, no. 1 (1994): 5–30.
6 C. Kennedy, S. Pincetl, and P. Bunje, "The Study of Urban Metabolism and Its Applications to Urban Planning and Design," *Environmental Pollution* 159, no. 8 (2011): 1965–1973.
7 Leon Battista Alberti, *On the Art of Building in Ten Books* (Boston: MIT Press, 1988), 114.
8 Mingtang Lei, Yongli Gao, and Xiaozhen Jiang, "Current Status and Strategic Planning of Sinkhole Collapses in China," in *Engineering Geology for Society and Territory—Volume 5*, ed. G. Lollino, A. Manconi, F. Guzzetti, M. Culshaw, P. Bobrowsky, and F. Luino (Heidelberg: Springer Cham, 2015), 529–533.
9 "The Sinkhole in Fukuoka," *Japan Times*, November 17, 2016, https://www.japantimes.co.jp/opinion/2016/11/17/editorials/the-sinkhole-in-fukuoka/; "Repaired Japan Sinkhole Sinks Again," *BBC News*, November 28, 2016, Asia, https://www.bbc.co.uk/news/world-asia-38129691.
10 World Bank, "Urban Population," *WorldBank.org*, 2017, https://data.worldbank.org/indicator/SP.URB.TOTL; also see Shinji Kaneko and Tomoyo Toyota, "Long-Term Urbanization and Land Subsidence in Asian Megacities: An Indicators System Approach," in *Groundwater and Subsurface Environments: Human Impacts in Asian Coastal Cities*, ed. Makoto Taniguchi (Tokyo: Springer Japan, 2011), 249–270; Antje Brunn, "The Environmental Impacts of Megacities on the Coast," in *Megacities and the Coast: Risk, Resilience and Transformation*, ed. Mark Pelling and Sophie Blackburn (London: Routledge, 2014), 22–69.
11 A. C. Oosthuizen and S. Richardson, "Sinkholes and Subsidence in South Africa," Council for Geoscience Report No. 2011-0010, Cape Town, 2011, https://www.geoscience.org.za/images/geohazard/Sinkholes.pdf.
12 Elsewhere, train service between Pretoria and Johannesburg has been repeatedly disrupted by sinkholes. See Thabo Molelekwa, "Sinkhole Havoc for Khutsong Community," *Health-E News*, June 9, 2017, https://www.health-e.org.za/2017/06/09/sinkhole-havoc-khutsong-community/.
13 Rodolfo G. Hermosilla, "The Guatemala City Sinkhole Collapses," *Carbonates and Evaporites* 27, no. 2 (2012): 103–107.

14. Robert Lionel Sherlock, *Man as a Geological Agent* (London: H. F. & G. Witherby, 1922), 14, also 132–133, https://archive.org/details/manasgeological a00sheriala.
15. Ian Douglas, *Cities: An Environmental History* (New York: I. B. Tauris, 2013).
16. Dora P. Crouch, *Water Management in Ancient Greek Cities* (New York: Oxford University Press, 1993); A. N. Angelakis, D. Koutsoyiannis, and G. Tchobanoglous, "Urban Wastewater and Stormwater Technologies in Ancient Greece," *Water Research* 39, no. 1 (2005): 210–220.
17. Carl A. Zimring, *Clean and White: A History of Environmental Racism in the United States* (New York: NYU Press, 2017), 56.
18. Heinz Hötzl and David Drew, "Industrial and Urban Produced Impacts," in *Karst Hydrogeology and Human Activities: Impacts, Consequences and Implications*, ed. David Drew (London: Routledge, 2017).
19. New York (State) Supreme Court Appellate Division, *Reports of Cases Heard and Determined in the Appellate Division of the Supreme Court of the State of New York* (Banks, 1905).
20. SCASD, "Memorial Field/Brief History," State College Area School District, 2018, accessed November 5, 2018, http://www.scasd.org/site/Default.aspx?PageID =1134.
21. "A Mill Creek Battle: Riparian Rights Threatened," *Philadelphia Inquirer*, November 19, 1890.
22. Anne Whiston Spirn, "Restoring Mill Creek: Landscape Literacy, Environmental Justice and City Planning and Design," *Landscape Research* 30, no. 3 (2005): 395–413.
23. Virginia Department of Health, *Virginia Health Bulletin* 18 (Richmond: Virginia Department of Health, 1926); Wisconsin State Board of Health, *State Board of Health Bulletin* (Madison, WI: Office of the State Board of Health, 1924); "Seventh Annual Report of the Commissioner of Health for the Commonwealth of Pennsylvania" (Harrisburg, PA: Division of Distribution of Biological Products, 1912).
24. Milton Joseph Rosenau, *Report on the Origin and Prevalence of Typhoid Fever in the District of Columbia* (Washington, DC: U.S. Government Printing Office, 1907).
25. S. P. Kingston, "Contamination of Water Supplies in Limestone Formation," *American Water Works Association* 35, no. 11 (1943): 1450–1456.
26. In Europe, this work is governed by the Urban Wastewater Treatment Directive and the Integrated Pollution Control and Prevention Directive. However, there is great variation among member states in the handling of stormwater. See Goncalo Moreira, *Assessment of Impact of Storm Water Overflows from Combined Waste Water Collecting Systems on Water Bodies (including the Marine Environment) in the 28 EU Member States* (Brussels: Milieu Ltd., 2016).
27. Ninad Bodhankar and B. Chatterjee, "Pollution of Limestone Aquifer Due to Urban Waste Disposal around Raipur, Madhya Pradesh, India," *Environmental Geology* 23, no. 3 (1994): 209–213; H. Elhatip, "The Influence of Karst Features on Environmental Studies in Turkey," *Environmental Geology* 31, no. 1 (1997): 27–33; Heinz Hötzl and David Drew, "Industrial and Urban Produced Impacts," in *Karst Hydrogeology and Human Activities: Impacts, Consequences and Implications*, ed. David Drew (London: Routledge, 2017); R. Brinkmann, M. Parise, and D. Dye, "Sinkhole Distribution in a Rapidly Developing Urban Environment:

Hillsborough County, Tampa Bay Area, Florida," *Engineering Geology* 99, no. 3 (2008): 169–184.

28 Examples include Hinsdale (Chicago), Upper Houghton County (Detroit), and Laurel Canyon (Los Angeles).

29 For example, Hurricane Maria flooded weakened and aging pipe systems, causing a sinkhole under a highway in Puerto Rico. See Laila Kearney, Nick Brown, and Hugh Bronstein, "In Puerto Rico, a Sinkhole of Rebuilding Struggles," *Reuters*, November 10, 2017, https://www.reuters.com/article/us-usa-puerto-rico-rebuild-insight/in-puerto-rico-a-sinkhole-of-rebuilding-struggles-idUSKBN1DA1IM.

30 SwissRe, "The Hidden Risks of Climate Change: An Increase in Property Damage from Soil Subsidence in Europe" (Zurich: Swiss Reinsurance Company Ltd., 2011); Miles Brignall, "Cracking Summer: UK Insurers Expect Rise in Subsidence Claims," *The Guardian*, August 17, 2018, Business, https://www.theguardian.com/business/2018/aug/17/cracking-summer-uk-insurers-expect-rise-subsidence-claims-heatwave.

31 Wanfang Zhou and Barry F. Beck, "Management and Mitigation of Sinkholes on Karst Lands: An Overview of Practical Applications," *Environmental Geology* 55, no. 4 (2008): 837–51.

32 In 1874, the Philadelphia *Public Ledger* called the Bureau of Indian Affairs a sinkhole, as republished in "Indian Department," *The Voice of Peace*, July 1874, 64. For a contemporary example, see: David Fahrenthold, "Sinkhole of Bureaucracy," *Washington Post*, March 22, 2014, https://www.washingtonpost.com/sf/national/2014/03/22/sinkhole-of-bureaucracy/

33 Robert M. Jameson, *Public Service Rates in Texas Cities* (Austin: University of Texas, 1914), 6.

34 Godfrey Nelson, "Taxes in New York and Costs Studied: Records of This and Various Other Northeastern States for 15 Years Compared," *New York Times*, February 17, 1946.

35 Daniel Webster Hoan, *Taxes and Tax Dodgers* (Chicago: Committee on Education and Research of the Socialist Party of America, 1933), 8.

36 Andrew Karvonen, *Politics of Urban Runoff: Nature, Technology, and the Sustainable City* (Cambridge, MA: MIT Press, 2011), 129–132.

37 "Sinkholes Make 5th Ave. Hazardous: Broadway, Too, from Fourteenth to Forty-Second Street Sadly in Need of Repair.," *New York Times*, January 29, 1911.

38 Governor Tom Wolf, "Harrisburg to Receive $1.65 Million in Funding for Sinkholes," news release, September 15, 2016, https://www.governor.pa.gov/harrisburg-to-receive-1-65-million-in-federal-funding-for-sinkhole-mitigation-project/; Donald Gilliland, "A Land of Sinkholes and Bursting Water Mains," *PennLive*, February 21, 2013, http://www.pennlive.com/midstate/index.ssf/2013/02/harrisburg_sinkhole_water_main.html.

39 Laura Osman, "You Call That a Sinkhole? Mayor Downplays Latest Rideau Street 'Degradation,'" *CBC News*, November 3, 2017, http://www.cbc.ca/news/canada/ottawa/sinkhole-hole-rideau-lrt-1.4385814.

40 Spencer Fleury, *Land Use Policy and Practice on Karst Terrains: Living on Limestone* (New York: Springer Science & Business Media, 2009).

41 Brinkmann, Parise, and Dye, "Sinkhole Distribution in a Rapidly Developing Urban Environment."

42   FLOIR, "Report on Review of 2010 Sinkhole Data Call," Florida Office of Insurance Regulation, November 8, 2010, https://www.floir.com/siteDocuments/Sinkholes/2010_Sinkhole_Data_Call_Report.pdf.
43   FLOIR, "Report on Review of the 2010 Sinkhole Data Call."
44   Florida Senate bill CS/CS/SB 552, or the Florida Springs and Aquifer Protection Act, was passed in 2016 to establish several water-monitoring programs as well as provide conservation and restoration efforts. This includes giving the Florida Department of Environmental Protection the ability to develop rules regarding groundwater withdrawals.
45   K. L. Eastman, A. M. Butler, and C. Lilly, "The Effect of Mandating Sinkhole Coverage in Florida Homeowners Insurance Policies," *CPCU Journal* 9 (1995): 165–176; E. Zisman, "The Florida Sinkhole Statute: Its Evolution, Impacts and Needed Improvements," *Carbonates and Evaporites* 28, no. 1/2 (2013): 95–102.
46   FEMA, "Total Coverage by Calendar Year," accessed July 12, 2015, http://www.fema.gov/total-coverage-calendar-year.
47   Leslie Scism, "Sinkhole Claims Threaten to Engulf Florida Insurers," *Wall Street Journal*, September 21, 2010, Business, http://www.wsj.com/articles/SB10001424052748703305004575504143442161532.
48   Susan Taylor Martin, "Citizens, Hit with $12.7 Million Verdict, Acted in 'Monumental Bad Faith,' Homeowner Says," *Tampa Bay Times*, March 14, 2018, https://www.tampabay.com/news/business/realestate/Citizens-hit-with-12-7-million-verdict-acted-in-monumental-bad-faith-homeowner-says_166336231.
49   Eric Glasser, "Hurricane Irma's Lingering Effect: Sinkholes," *WTSP 10NEWS*, accessed December 24, 2017, http://www.wtsp.com/weather/irma/hurricane-irmas-lingering-effect-sinkholes/477358541.
50   Andrew G. Simpson, "What's Next: The Florida Sinkhole Insurance Fund?," *Insurance Journal*, November 1, 2010, https://www.insurancejournal.com/magazines/mag-features/2010/11/01/160368.htm.
51   Leslie Scism, "Why Florida Is Largely Insured by Companies You've Never Heard Of," *Wall Street Journal*, September 7, 2017, Markets, https://www.wsj.com/articles/why-florida-is-largely-insured-by-companies-youve-never-heard-of-1504782003.
52   "Second Sinkhole Appears. Does Your Insurance Cover Sinkholes?," *Christian Science Monitor*, March 5, 2013, https://www.csmonitor.com/Business/2013/0305/Second-sinkhole-appears.-Does-your-insurance-cover-sinkholes.
53   Lloyd Dunkelberger News Service of Florida, "Florida Adds Hurricane Reinsurance for 'Healthy' Catastrophe Fund," *OrlandoSentinel.com*, May 23, 2017, http://www.orlandosentinel.com/business/os-nsf-florida-hurricane-insurance-20170523-story.html; Paul J. Davies, "Hurricane Irma Tests Risk-Reward of Catastrophe-Bond Market," *Wall Street Journal*, September 8, 2017, Markets, https://www.wsj.com/articles/hurricane-irma-tests-risk-reward-of-catastrophe-bond-market-1504895037.
54   Laïla Smith, "The Murky Waters of the Second Wave of Neoliberalism: Corporatization as a Service Delivery Model in Cape Town," *Geoforum* 35, no. 3 (2004): 375–393; Jamie Peck and Adam Tickell, "Neoliberalizing Space," *Antipode* 34, no. 3 (2002): 380–404.
55   Aileen Perilla, "Citizens Insurance 'Incredibly Strong' despite Nearly $2 Billion in Hurricane Irma Claims—Orlando Sentinel," *Orlando Sentinel*, September 26,

2018, https://www.orlandosentinel.com/news/politics/political-pulse/os-hurricane-irma-insurance-claims-20180926-story.html.

56 M. J. Thornbush, "Part 1: Contemporary Challenges and Current Solutions in Sinkhole Occurrence and Mitigation," *Journal of Geology & Geophysics* 6, no. 3 (2017); Christopher Balogh, "Why Are There So Many Sinkholes in Florida?," *The Atlantic*, August 22, 2014, https://www.theatlantic.com/technology/archive/2014/08/why-are-there-so-many-sinkholes-in-florida/378869/; Sonja Alexander, "NASA Radar Demonstrates Ability to Foresee Sinkholes," NASA news release, March 6, 2014, http://www.nasa.gov/press/2014/march/nasa-radar-demonstrates-ability-to-foresee-sinkholes.

57 Jeff Libby and Erin Ailworth, "Avoid Sinkhole Tests, Experts Say," *Orlando Sentinel*, January 15, 2005, https://www.orlandosentinel.com/orl-sinkhole011205-story.html; Larry D. Croom, "Experts Agree That Sinkhole-Damaged Home on McLawren Terrace Must Be Torn Down," *Villages News*, July 6, 2018, https://www.villages-news.com/experts-agree-that-sinkhole-damaged-home-on-mclawren-terrace-must-be-torn-down/.

58 Holly P. Jones, David G. Hole, and Erika S. Zavaleta, "Harnessing Nature to Help People Adapt to Climate Change," *Nature Climate Change* 2, no. 7 (2012): 504.

59 Ryan Gunderson, Diana Stuart, and Brian Petersen, "The Political Economy of Geoengineering as Plan B: Technological Rationality, Moral Hazard, and New Technology," *New Political Economy* 24, no. 5 (2018): 696–715.

60 Gunderson, Stuart, and Petersen, "The Political Economy of Geoengineering," 4.

61 Kenneth Hewitt, "The Idea of Calamity in a Technocratic Age," in *Interpretations of Calamity*, ed. Kenneth Hewitt (Boston: Allen & Unwin, 1983), 3–32; Kenneth Hewitt, *Regions of Risk: A Geographical Introduction to Disasters* (New York: Longman, 1997); William R. Freudenburg, Robert Gramling, Shirley Laska, and Kai T. Erikson, "Organizing Hazards, Engineering Disasters?," *Social Forces* 87, no. 2 (2008): 1015–1038; Benjamin Wisner, Piers Blaikie, Terry Cannon, and Ian Davis, *At Risk: Natural Hazards, People's Vulnerability and Disasters* (Oxford: Routledge, 1994).

62 UNESCO and International Hydrological Programme, "Proposal for the Establishment of the Land Subsidence International Initiative (LaSII), 23rd Session of the Intergovernmental Council," Paris, June 11–15, 2018, 1, https://en.unesco.org/sites/default/files/ic-xiii_ref_5_land_subsidence.pdf.

Chapter 3 The Logistical Nightmare of Trash and Urban Nature

1 Angela Giuffrida, "Rome Told to Sort Out Rubbish Crisis or Face EU Sanctions," *The Guardian*, January 10, 2018, http://www.theguardian.com/world/2018/jan/10/rome-rubbish-crisis-italy-eu-sanctions; Elisabetta Povoledo, "EU Criticizes Italy over Trash Crisis in Naples," *New York Times*, January 15, 2008, Europe, https://www.nytimes.com/2008/01/15/world/europe/15iht-italy.4.9237559.html.

2 Stuart Chase, *Waste and the Machine Age* (New York: League for Industrial Democracy, 1931), 48.

3 Waste Management, Inc., "Waste Management Announces Fourth Quarter and Full-Year 2020 Earnings," February 18, 2021, http://investors.wm.com/news-releases/news-release-details/waste-management-announces-fourth-quarter-and-full-year-2020

4 Michael Thompson, *Rubbish Theory: The Creation and Destruction of Value* (Oxford: Oxford University Press, 1979).
5 Of course, a problem is that "virgin" plastic made from petrochemicals is cheaper. This undermines the effectiveness of recycling.
6 Jen Baggs, "International Trade in Hazardous Waste," *Review of International Economics* 17, no. 1 (2009): 1–16; R. Scott Frey, "The Export of Hazardous Industries to the Peripheral Zones of the World-System," *Journal of Developing Societies* 14, no. 1 (1998): 66–81; R. Scott Frey, "The Transfer of Core-Based Hazardous Production Processes to the Export Processing Zones of the Periphery: The Maquiladora Centers of Northern Mexico," *Journal of World-Systems Research* 9, no. 2 (2003): 317–354.
7 See Basil Horen, "Fragmented Coherence: Solid Waste Management in Colombo," *International Journal of Urban and Regional Research* 28, no. 4 (2004): 757–773; Mary Lawhon, "Dumping Ground or Country-in-Transition? Discourses of e-Waste in South Africa," *Environment and Planning C: Government and Policy* 31, no. 4 (2013): 700–715; Camilla Louise Bjerkli, "Governance on the Ground: A Study of Solid Waste Management in Addis Ababa, Ethiopia: Solid Waste Management in Addis Ababa," *International Journal of Urban and Regional Research* 37, no. 4 (2013): 1273–1287.
8 Raquel Pinderhughes, *Alternative Urban Futures: Planning for Sustainable Development in Cities throughout the World* (New York: Rowman & Littlefield, 2004).
9 Sandra Cointreau-Levine, *Private Sector Participation in Municipal Solid Waste Services in Developing Countries*, vol. 1 (Washington, DC: World Bank, Urban Management Programme, 1994), http://inswa.or.id/wp-content/uploads/2012/07/Private-Sector-Participation-in-Municipal-Solid-Waste-Services-in-Developing-Countries3.pdf; Johan Post, Jaap Broekema, and Nelson Obirih-Opareh, "Trial and Error in Privatisation: Experiences in Urban Solid Waste Collection in Accra (Ghana) and Hyderabad (India)," *Urban Studies* 40, no. 4 (2003): 835–852.
10 Edward Humes, *Garbology: Our Dirty Love Affair with Trash* (New York: Avery Trade, 2013); Adam Diamond, "What a Waste: Municipal Refuse Reform and a Century of Solid-Waste Management in Los Angeles," *Southern California Quarterly* 88, no. 3 (2006): 339–365.
11 Juvenal, *The Sixteen Satires* (London: Penguin, 2004), 3.269–277.
12 Hans Y. Tammemagi, *The Waste Crisis: Landfills, Incinerators, and the Search for a Sustainable Future* (New York: Oxford University Press, 1999).
13 Corbyn Morris, *Observations on the Past Growth and Present State of the City of London* (London, 1751).
14 It is important to note the role of inequality and social marginalization in creating this workforce and how concepts of "clean" are also used to create categories of people to do this work. David C. Wilson, Costas Velis, and Chris Cheeseman, "Role of Informal Sector Recycling in Waste Management in Developing Countries," *Habitat International* 30, no. 4 (2006): 797–808; also see Carl A. Zimring, *Clean and White: A History of Environmental Racism in the United States* (New York: NYU Press, 2017).
15 Mark Jayne, *Cities and Consumption* (London: Routledge, 2005).
16 Marina Fischer-Kowalski and Christof Amann, "Beyond IPAT and Kuznets Curves: Globalization as a Vital Factor in Analysing the Environmental Impact of Socio-Economic Metabolism," *Population and Environment* 23, no. 1 (2001):

7–47; Richard York, Eugene A. Rosa, and Thomas Dietz, "STIRPAT, IPAT and ImPACT: Analytic Tools for Unpacking the Driving Forces of Environmental Impacts," *Ecological Economics* 46, no. 3 (2003): 351–365.
17. Diane Sicotte, *From Workshop to Waste Magnet: Environmental Inequality in the Philadelphia Region* (New Brunswick, NJ: Rutgers University Press, 2016).
18. Jerome I. Hodos, *Second Cities* (Philadelphia: Temple University Press, 2011).
19. "Metropolitan Paper Recycling—Company Overview," accessed December 20, 2016, http://www.metropaperrecycling.com/comp_about.html.
20. In February 2015, Veolia and the firm Lockwood, Andrews & Newnam (LAN) were hired by the city of Flint to assess drinking water quality.
21. David Hall and Tue Anh Nguyen, "Waste Management in Europe: Companies, Structure and Employment," Public Services International Research Unit (PSIRU), University of Greenwich, 2012.
22. Germà Bel, Elbert Dijkgraaf, Xavier Fageda, and Raymond Gradus, "Similar Problems, Different Solutions: Comparing Refuse Collection in the Netherlands and Spain," *Public Administration* 88, no. 2 (2010): 479–495.
23. Eugene McCann, "Urban Policy Mobilities and Global Circuits of Knowledge: Toward a Research Agenda," *Annals of the Association of American Geographers* 101, no. 1 (2011): 107–130; Jamie Peck and Nik Theodore, "Exporting Workfare/Importing Welfare-to-Work: Exploring the Politics of Third Way Policy Transfer," *Political Geography* 20, no. 4 (2001): 427–460.
24. Tim Bunnell and Diganta Das, "Urban Pulse—A Geography of Serial Seduction: Urban Policy Transfer from Kuala Lumpur to Hyderabad," *Urban Geography* 31, no. 3 (April 1, 2010): 277–284.
25. Wilson, Velis, and Cheeseman, "Role of Informal Sector Recycling." See also Faranak Miraftab, "Neoliberalism and Casualization of Public Sector Services: The Case of Waste Collection Services in Cape Town, South Africa," *International Journal of Urban and Regional Research* 28, no. 4 (2004): 874–892; Franklin Obeng-Odoom, "Green Neoliberalism: Recycling and Sustainable Urban Development in Sekondi-Takoradi," *Habitat International* 41 (2014): 129–134; Martin Oteng-Ababio, "Private Sector Involvement in Solid Waste Management in the Greater Accra Metropolitan Area in Ghana," *Waste Management & Research* 28, no. 4 (2010): 322–329.
26. Ariella Cohen, "Development atop a City Dump? Policy Fiasco Comes Back to Haunt Upper 9th Ward," *Louisiana Weekly*, January 17, 2012, http://www.louisianaweekly.com/development-atop-a-city-dump-policy-fiasco-comes-back-to-haunt-upper-9th-ward/.
27. Derek Kellenberg, "Consumer Waste Backhauling, and Pollution Havens," *Journal of Applied Economics* 13, no. 2 (2010): 283–304; Derek Kellenberg, "Trading Wastes," *Journal of Environmental Economics and Management* 64, no. 1 (2012): 68–87.
28. Arthur P. J. Mol, *Globalization and Environmental Reform: The Ecological Modernization of the Global Economy* (Boston: MIT Press, 2003).
29. Adam Weinberg, Allan Schnaiberg, and Kenneth A. Gould, "Recycling: Conserving Resources or Accelerating the Treadmill of Production?," *Advances in Human Ecology* 4 (1995): 173–205; Adam S. Weinberg, David N. Pellow, and Allan Schnaiberg, *Urban Recycling and the Search for Sustainable Community Development* (Princeton, NJ: Princeton University Press, 2000).

30 In addition to trash, Jorgenson and Clark found an intensification of carbon emissions from 1960 to 2005 alongside gross domestic product (GDP) increases. Again, outcomes vary between developed and less developed countries. See Andrew K. Jorgenson and Brett Clark, "Are the Economy and the Environment Decoupling? A Comparative International Study, 1960–2005," *American Journal of Sociology* 118, no. 1 (2012): 1–44.
31 Massimiliano Mazzanti and Roberto Zoboli, "Delinking and Environmental Kuznets Curves for Waste Indicators in Europe," *Environmental Sciences* 2, no. 4 (2005): 409–425; Massimiliano Mazzanti and Roberto Zoboli, "Waste Generation, Waste Disposal and Policy Effectiveness: Evidence on Decoupling from the European Union," *Resources, Conservation and Recycling* 52, no. 10 (2008): 1221–1234.
32 Amandine Gnonlonfin, Yusuf Kocoglu, and Nicolas Péridy, "Municipal Solid Waste and Development: The Environmental Kuznets Curve Evidence for Mediterranean Countries," *Région et Développement*, no. 45 (2017): 113–130.
33 Marina Alberti, John M. Marzluff, Eric Shulenberger, Gordon Bradley, Clare Ryan, and Craig Zumbrunnen, "Integrating Humans into Ecology: Opportunities and Challenges for Studying Urban Ecosystems," *BioScience* 53, no. 12 (2003): 1169–1179; X. M. Li, R. B. Xiao, S. H. Yuan, J. An. Chen, and J. X. Zhou, "Urban Total Ecological Footprint Forecasting by Using Radial Basis Function Neural Network: A Case Study of Wuhan City, China," *Ecological Indicators* 10, no. 2 (2010): 241–248.
34 Andrew K. Jorgenson, "Consumption and Environmental Degradation: A Cross-National Analysis of the Ecological Footprint," *Social Problems* 50, no. 3 (2003): 374–394; Andrew K. Jorgenson and Brett Clark, "The Economy, Military, and Ecologically Unequal Exchange Relationships in Comparative Perspective: A Panel Study of the Ecological Footprints of Nations, 1975–2000," *Social Problems* 56, no. 4 (2009): 621–646; Thomas J. White, "Sharing Resources: The Global Distribution of the Ecological Footprint," *Ecological Economics* 64, no. 2 (2007): 402–410.
35 Laïla Smith, "The Murky Waters of the Second Wave of Neoliberalism: Corporatization as a Service Delivery Model in Cape Town," *Geoforum* 35, no. 3 (2004): 375–393.
36 Smith, "The Murky Waters," 382; also see Margaret Walls, Molly Macauley, and Soren Anderson, "Private Markets, Contracts, and Government Provision: What Explains the Organization of Local Waste and Recycling Markets?," *Urban Affairs Review* 40, no. 5 (2005): 590–613.
37 Adrian Parr, *Hijacking Sustainability* (Boston: MIT Press, 2009); also see Miraftab, "Neoliberalism and Casualization"; Oteng-Ababio, "Private Sector Involvement in Solid Waste Management"; Wilson, Velis, and Cheeseman, "Role of Informal Sector Recycling."
38 "Actividades de Construcción y Servicios, S.A.—Construction," accessed November 16, 2016, https://www.grupoacs.com/activities/infrastructure/; http://www.grupoacs.com/index.php/en/c/activities_construction.
39 Hall and Nguyen, "Waste Management in Europe."
40 Kevin F. Gotham, "The Secondary Circuit of Capital Reconsidered: Globalization and the U.S. Real Estate Sector," *American Journal of Sociology* 112, no. 1 (2006): 231–275; Paul L. Knox, "World Cities in a World-System," in *World Cities in a*

World-System, ed. Paul L. Knox and Peter J. Taylor (Cambridge: Cambridge University Press, 1995), 3–20; David Parker, *Global Real Estate Investment Trusts: People, Process and Management*, vol. 33 (Oxford: Wiley-Blackwell, 2012); Peter J. Taylor, "Specification of the World City Network," *Geographical Analysis* 33, no. 2 (2001): 181–194.

41 Lois W. Morton, Yu-Che Chen, and Ricardo S. Morsea, "Small Town Civic Structure and Interlocal Collaboration for Public Services," *City & Community* 7, no. 1 (2008): 45–60.

42 Reuters Staff, "U.S. Asks China Not to Implement Ban on Foreign Garbage," *Reuters*, March 23, 2018, https://www.reuters.com/article/us-china-environment-usa/u-s-asks-china-not-to-implement-ban-on-foreign-garbage-idUSKBN1GZ2WI; David Stanway, "China Trash Town's Cleanup Bolstered by Import Ban," *Reuters*, January 24, 2018, https://www.reuters.com/article/us-china-environment-waste-insight/china-trash-towns-cleanup-bolstered-by-import-ban-idUSKBN1FD043; Committee on Technical Barriers to Trade, "China's Import Ban on Solid Waste," World Trade Organization, July 18, 2017, https://docs.wto.org/dol2fe/Pages/SS/directdoc.aspx?filename=q:/G/TBTN17/CHN1211.pdf.

43 "The Waste Group—Company History," accessed January 3, 2016, http://www.wastegroup.co.za/our-company/history/.

44 Albert S. Fu, "Neoliberalism, Logistics and the Treadmill of Production in Metropolitan Waste Management: A Case of Turkish Firms," *Urban Studies* 53, no. 10 (2016): 2099–2117.

45 Jean-Luc Guffond and Gilbert Leconte, "Developing Construction Logistics Management: The French Experience," *Construction Management and Economics* 18, no. 6 (2000): 679–687.

46 Craig Martin, "Shipping Container Mobilities, Seamless Compatibility, and the Global Surface of Logistical Integration," *Environment and Planning A* 45, no. 5 (2013): 1021–1036; Cynthia Negrey, Jeffery L. Osgood, and Frank Goetzke, "One Package at a Time: The Distributive World City," *International Journal of Urban and Regional Research* 35, no. 4 (2011): 812–831. Cross-sectoral strategies are often deployed to increase efficiency in other areas as well. For instance, see Rachael Shwom and Analena Bruce, "U.S. Non-Governmental Organizations' Cross-Sectoral Entrepreneurial Strategies in Energy Efficiency," *Regional Environmental Change* 18, no. 5 (2018): 1309–1321.

47 Mol, *Globalization and Environmental Reform*; Anne Scheinberg and Arthur P. J. Mol, "Multiple Modernities: Transitional Bulgaria and the Ecological Modernisation of Solid Waste Management," *Environment and Planning C, Government and Policy* 28, no. 1 (2010): 18.

48 Markus Hesse and Jean-Paul Rodrigue, "The Transport Geography of Logistics and Freight Distribution," *Journal of Transport Geography* 12, no. 3 (2004): 171–184; Markus Hesse, *The City as a Terminal: The Urban Context of Logistics and Freight Transport* (London: Routledge, 2008); William I. Robinson, "Remapping Development in Light of Globalisation: From a Territorial to a Social Cartography," *Third World Quarterly* 23, no. 6 (2002): 1047–1071; Jean-Paul Rodrigue, "Globalization and the Synchronization of Transport Terminals," *Journal of Transport Geography* 7, no. 4 (1999): 255–261; Todd Swanstrom, "Beyond Economism: Urban Political Economy and the Postmodern Challenge," *Journal of Urban Affairs* 15, no. 1 (1993): 55–78.

49 Murat Arsel, "Reflexive Developmentalism? Toward an Environmental Critique of Modernization," in *Environmentalism in Turkey: Between Democracy and Development?*, ed. Fikret Adaman and Murat Arsel (Aldershot, UK: Ashgate Publishing, 2005), 15–34.

50 Karen Bakker, "The 'Commons' versus the 'Commodity': Alter-Globalization, Anti-Privatization and the Human Right to Water in the Global South," *Antipode* 39, no. 3 (2007): 430–455; James McCarthy, "Privatizing Conditions of Production: Trade Agreements as Neoliberal Environmental Governance," *Geoforum* 35, no. 3 (2004): 327–341; Harold A. Perkins, "Out from the (Green) Shadow? Neoliberal Hegemony through the Market Logic of Shared Urban Environmental Governance," *Political Geography* 28, no. 7 (2009): 395–405.

51 Richard Batley, "Public-Private Relationships and Performance in Service Provision," *Urban Studies* 33, no. 4–5 (1996): 723–751.

52 Investment Support and Promotion Agency, *Turkish Environmental Technologies and Renewable Energy Industry Report* (Ankara: Republic of Turkey Prime Ministry, 2010), accessed February 23, 2015, http://www.invest.gov.tr/en-US/infocenter/publications/Documents/ENVIRONMENTAL.TECH.RENEWABLE.ENERGY.PDF.

53 Neil Brenner and Nik Theodore, "Cities and the Geographies of 'Actually Existing Neoliberalism,'" *Antipode* 34, no. 3 (2002): 349–379; Fu, "Neoliberalism, Logistics and the Treadmill of Production."

54 K. A. Gould, D. N. Pellow, and A. Schnaiberg, "Interrogating the Treadmill of Production," *Organization & Environment* 17, no. 3 (2004): 296–316; Jorgenson and Clark, "The Economy, Military, and Ecologically Unequal"; Allan Schnaiberg, *The Environment, from Surplus to Scarcity* (New York: Oxford University Press, 1980).

55 Schnaiberg, *The Environment*, 247.

56 A. Schnaiberg, D. N. Pellow, and A. Weinberg, "The Treadmill of Production and the Environmental State," in *The Environmental State under Pressure*, ed. A. P. J. Mol and F. H. Buttel, vol. 10, Research in Social Problems and Public Policy (Bingley, UK: Emerald Publishing, 2002), 15–32.

57 Ed Brown, Ben Derudder, Christof Parnreiter, Wim Pelupessy, Peter J. Taylor, and Frank Witlox, "World City Networks and Global Commodity Chains: Towards a World-Systems' Integration," *Global Networks* 10, no. 1 (2010): 12–34; also see Ben Derudder, "On Conceptual Confusion in Empirical Analyses of a Transnational Urban Network," *Urban Studies* 43, no. 11 (2006): 2027–2046; Paul L. Knox and Peter J. Taylor, eds., *World Cities in a World-System* (Cambridge: Cambridge University Press, 1995).

58 Maarten de Kadt, "Solid Waste Management at a Crossroads: Recycling on the Treadmill of Production," *Capitalism Nature Socialism* 10, no. 3 (1999): 131–160; Weinberg, Schnaiberg, and Gould, "Recycling."

59 Stephen Anderson, Julian Allen, and Michael Browne, "Urban Logistics—How Can It Meet Policy Makers' Sustainability Objectives?," *Journal of Transport Geography* 13, no. 1 (2005): 71–81; Fu, "Neoliberalism, Logistics and the Treadmill of Production."

60 Anderson, Allen, and Browne, "Urban Logistics"; Jesús Muñuzuri, Juan Larrañeta, Luis Onieva, and Pablo Cortés, "Solutions Applicable by Local Administrations for Urban Logistics Improvement," *Cities* 22, no. 1 (2005): 15–28.

61 Frederick H. Buttel, "Social Institutions and Environmental Change," in *The International Handbook of Environmental Sociology*, ed. Michael Redclift and Graham Woodgate (Northampton, MA: Elgar, 1997), 39. See also Frederick H. Buttel, "Ecological Modernization as Social Theory," *Geoforum* 31, no. 1 (2000): 57–65.

62 Harvey Molotch, "The City as a Growth Machine: Toward a Political Economy of Place," *American Journal of Sociology* 82, no. 2 (1976): 309–332; Harvey Molotch and John Logan, "Tensions in the Growth Machine: Overcoming Resistance to Value-Free Development," *Social Problems* 31, no. 5 (1984): 483–499; John Logan and Harvey Molotch, *Urban Fortunes: The Political Economy of Place* (Berkeley: University of California Press, 1987).

63 Albert S. Fu, "Connecting Urban and Environmental Catastrophe: Linking Natural Disaster, the Built Environment, and Capitalism," *Environmental Sociology* 2, no. 4 (2016): 365–374.

64 Kevin R. Cox, "Questions of Abstraction in Studies in the New Urban Politics," *Journal of Urban Affairs* 13, no. 3 (1991): 267–280; Mark Schneider and Paul Teske, "The Antigrowth Entrepreneur: Challenging the 'Equilibrium' of the Growth Machine," *Journal of Politics* 55, no. 03 (1993): 720–736.

65 Stephen G. Bunker, "How Ecologically Uneven Developments Put the Spin on the Treadmill of Production," *Organization & Environment* 18, no. 1 (2005): 38–54; also see Immanuel Wallerstein, *Historical Capitalism: With Capitalist Civilization* (London: Verso Books, 2011).

66 James Rice, "The Transnational Organization of Production and Uneven Environmental Degradation and Change in the World Economy," *International Journal of Comparative Sociology* 50, no. 3–4 (2009): 215–236.

67 Liat Clark, "Sweden to Import 800,000 Tonnes of Trash to Burn for Energy," *Wired UK*, October 28, 2012, http://www.wired.co.uk/article/sweden-imports-garbage-for-energy; Bonnie Kavoussi, "Sweden Must Import Trash for Energy Conversion Because Its Recycling Program Is So Successful," *Huffington Post*, September 12, 2012, Business, http://www.huffingtonpost.com/2012/09/12/sweden-imports-trash_n_1876746.html; Hazel Sheffield, "Sweden's Recycling Is So Revolutionary the Country Has Run Out of Rubbish," *The Independent*, December 8, 2016, http://www.independent.co.uk/environment/sweden-s-recycling-is-so-revolutionary-the-country-has-run-out-of-rubbish-a7462976.html.

68 Carrie Arnold, "Is Sustainable Trash-Burning a Load of Rubbish?," *Smithsonian*, August 1, 2016, http://www.smithsonianmag.com/science-nature/burning-trash-solution-our-garbage-woes-or-are-advocates-just-blowing-smoke-180959924/.

69 Howard Fischer, "Is Garbage a Renewable Energy Source like Wind and Solar?," *Arizona Daily Sun*, November 4, 2015, https://azdailysun.com/info/is-garbage-a-renewable-energy-source-like-wind-and-solar/article_3f65ad20-ae51-53b3-bd73-59fe4243 50d9.html.

70 IEA, "Will Energy from Waste Become the Key Form of Bioenergy in Asia?—Analysis," International Energy Association, January 10, 2019, https://www.iea.org/articles/will-energy-from-waste-become-the-key-form-of-bioenergy-in-asia.

71 CEWEP, "Bottom Ash Fact Sheet," Confederation of European Waste-to-Energy Plants, accessed September 5, 2021, https://www.cewep.eu/wp-content/uploads/2017/09/FINAL-Bottom-Ash-factsheet.pdf.

72 Ciarán J. Lynn, Gurmel S. Ghataora, and Ravindra K. Dhir OBE, "Municipal Incinerated Bottom Ash (MIBA) Characteristics and Potential for Use in Road

Pavements," *International Journal of Pavement Research and Technology* 10, no. 2 (2017): 185–201.
73   Weinberg, Pellow, and Schnaiberg, *Urban Recycling*; Weinberg, Schnaiberg, and Gould, "Recycling."
74   De Kadt, "Solid Waste Management at a Crossroads"; also see Martin Hingley, Adam Lindgreen, David B. Grant, and Charles Kane, "Using Fourth-Party Logistics Management to Improve Horizontal Collaboration among Grocery Retailers," *Supply Chain Management: An International Journal* 16, no. 5 (2011): 316–327.
75   Aneeta Mary Joseph, Ruben Snellings, Philip Van den Heede, Stijn Matthys, and Nele De Belier, "The Use of Municipal Solid Waste Incineration Ash in Various Building Materials: A Belgian Point of View," *Materials* 11, no. 1 (2018), https://www.ncbi.nlm.nih.gov/pmc/articles/PMC5793639/.
76   M. B. Ali, R. Saidur, and M. S. Hossain, "A Review on Emission Analysis in Cement Industries," *Renewable and Sustainable Energy Reviews* 15, no. 5 (2011): 2252–2261.
77   Suez Environment Company, "Reference Document 2015," French Financial Markets Authority and Suez Group, April 4, 2016, 107, https://www.suez.com/-/media/suez-global/files/publication-docs/pdf-english/reference-document-2015.pdf.
78   Suez Environment Company, "Reference Document 2015," French Financial Markets Authority and Suez Group, April 5, 2018, https://www.suez.com/-/media/suez-global/files/publication-docs/pdf-english/suez-reference-document-2018-en-compressed.pdf.
79   GDF Suez, "Our History in China," accessed, September 5, 2021, https://www.engie.cn/en/gdf-suez-in-china/our-history-in-china/.
80   GDF Suez, "GDF Suez Energy International Facts and Figures," accessed March 1, 2017, https://www.engie.com/wp-content/uploads/2014/09/gdf-suez-energy-international-march-2015.pdf.
81   Emmanuel, "Sustainable Waste Management—Cameroon Targets 100 MW of Energy from Household Waste," *Cameroon Today*, September 13, 2011, http://news.cameroon-today.com/sustainable-waste-management-cameroon-targets-100-mw-of-energy-from-household-waste/7641/.
82   Logan and Molotch, *Urban Fortunes*; Molotch, "The City as a Growth Machine"; Schnaiberg, Pellow, and Weinberg, "The Treadmill of Production and the Environmental State."
83   Richard York, Eugene A. Rosa, and Thomas Dietz, "Footprints on the Earth: The Environmental Consequences of Modernity," *American Sociological Review* 68, no. 2 (April 2003): 279–300.
84   David Naguib Pellow, *Garbage Wars: The Struggle for Environmental Justice in Chicago* (Boston: MIT Press, 2003).
85   "Privatisation: Solid Waste Management Staff on Strike," *Express Tribune*, January 9, 2013, http://tribune.com.pk/story/491110/privatisation-solid-waste-management-staff-on-strike/.
86   Tom Burridge, "Madrid: Rubbish Piles up on Streets as Cleaners Strike," *BBC News*, November 14, 2013, Europe, http://www.bbc.com/news/world-europe-24941737.
87   La Tercera, "Conoce El Perfil de Las Empresas Que Recolectan y Tratan La Basura de La Región Metropolitana," *La Tercera*, August 2, 2013, http://www.latercera

.com/noticia/conoce-el-perfil-de-las-empresas-que-recolectan-y-tratan-la-basura-de-la-region-metropolitana/.
88. Sarah Hodgson, "Protest Shuts A548 in Flintshire in Row over £800m Waste Incinerator Plant," *North Wales Daily Post*, October 19, 2017, http://www.dailypost.co.uk/news/north-wales-news/protest-shuts-a548-flintshire-row-13786329.
89. Petra Kuppinger, "Crushed? Cairo's Garbage Collectors and Neoliberal Urban Politics," *Journal of Urban Affairs* 36, no. s2 (2014): 621–633; Sarah A. Moore, "The Excess of Modernity: Garbage Politics in Oaxaca, Mexico," *The Professional Geographer* 61, no. 4 (2009): 426–437.
90. Gary Fetter and Terry Rakes, "Incorporating Recycling into Post-Disaster Debris Disposal," *Socio-Economic Planning Sciences* 46, no. 1 (2012): 14–22.
91. Charlotte Brown, Mark Milke, and Erica Seville, "Disaster Waste Management: A Review Article," *Waste Management* 31, no. 6 (2011): 1085–1098.
92. U.S. Army Corps of Engineers, "Northern California Wildfire Response," accessed July 15, 2018, http://www.spk.usace.army.mil/CAwildfire/.
93. Paul Payne, "More than 200,000 Tons of Debris Cleaned up from Santa Rosa Fires," *Santa Rosa Press Democrat*, November 30, 2017, http://www.pressdemocrat.com/news/7695881-181/more-than-200000-tons-of; Emily Turner, "Wildfire Debris Causes Problems at Sonoma County Landfill," *CBS SF BayArea*, December 8, 2017, http://sanfrancisco.cbslocal.com/2017/12/08/wildfire-debris-gridlock-sonoma-landfill/; Michael Crowe, "New Construction Outpacing Wildfire Debris at Sevier Co. Landfill," *WBIR.com*, November 15, 2017, https://www.wbir.com/article/news/local/new-construction-outpacing-wildfire-debris-at-sevier-co-landfill/51-492153955; The Canadian Press, "Wildfire Waste a Challenge for Fort McMurray Landfill," *CBC News*, July 10, 2016, http://www.cbc.ca/news/canada/edmonton/wildfire-waste-a-challenge-for-fort-mcmurray-landfill-1.3672588.
94. As an interesting aside, the Fluor Corporation also sells carbon capture technologies for power plants. In other words, it sells climate change "mitigation" strategies. However, as mentioned in earlier chapters, this only encourages the ongoing use of fossil fuels.

Chapter 4   Fire, the Wildland–Urban Interface, and Feedback Loops

1. Andrew C. Scott, *Burning Planet: The Story of Fire Through Time*. (New York: Oxford University Press, 2018); Utku Balaban and Albert S. Fu, "Politics of Urban Development and Wildfires in California and Turkey," *Environment and Planning A* 46, no. 4 (2014): 820–836; FAO, "Community-Based Fire Management: Case Studies from China, The Gambia, Honduras, India, the Lao People's Democratic Republic and Turkey," (Bangkok, Thailand: Food and Agriculture Organization of the United Nations, 2003); FAO, "Wildfire Prevention in the Mediterranean," Position Paper, 2011; J. G. Pausas and S. Fernández-Muñoz, "Fire Regime Changes in the Western Mediterranean Basin: From Fuel-Limited to Drought-Driven Fire Regime," *Climatic Change* 110, no. 1–2 (2012): 215–226.
2. John Twigg, Nicola Christie, James Haworth, Emmanuel Osuteye, and Artemis Skarlatidou, "Improved Methods for Fire Risk Assessment in Low-Income and Informal Settlements," *International Journal of Environmental Research and Public Health* 14, no. 2 (2017): 139.

3 Jeffrey Jones, "Syncrude Canada Oil Sands Operation Shut Down, Staff Removed, Due to Fort McMurray Fire," *The Globe and Mail*, May 7, 2016, https://www.theglobeandmail.com/report-on-business/industry-news/energy-and-resources/syncrude-canada-oil-sands-shut-down-staff-removed-due-to-fire/article29932912/.

4 A. Malcolm Gill, Scott L. Stephens, and Geoffrey J. Cary, "The Worldwide 'Wildfire' Problem," *Ecological Applications* 23, no. 2 (2013): 438–454; William E. Mell, Samuel L. Manzello, Alexander Maranghides, David Butry, and Ronald G. Rehm, "The Wildland–Urban Interface Fire Problem: Current Approaches and Research Needs," *International Journal of Wildland Fire* 19, no. 2 (2010): 238–251.

5 Meg A. Krawchuk, Max A. Moritz, Marc-André Parisien, Jeff Van Dorn, and Katharine Hayhoe, "Global Pyrogeography: The Current and Future Distribution of Wildfire," *PLoS One* 4, no. 4 (2009): e5102; Yongqiang Liu, John Stanturf, and Scott Goodrick, "Trends in Global Wildfire Potential in a Changing Climate," *Forest Ecology and Management* 259, no. 4 (2010): 685–697; A. L. Westerling and B. P. Bryant, "Climate Change and Wildfire in California," *Climatic Change* 87, no. S1 (2007): 231–249.

6 Ricardo Vélez, "Mediterranean Forest Fires: A Regional Perspective," *Unasylva: An International Journal of Forestry and Forest Industries*, 41, no. 3 (1990), http://www.fao.org/docrep/t9500e/t9500e02.htm#mediterranean%20forest%20fires:%20a%20regional%20perspective; Raquel Pinderhughes, *Alternative Urban Futures: Planning for Sustainable Development in Cities throughout the World* (New York: Rowman & Littlefield, 2004); Marty Ahrens, "Local Fire Department Responses to Wildfires in the U.S.: National Estimates Based on 2004–08 Data," *Fire Safety Science* 10 (2011): 1389–1400.

7 William E. Rees, "Ecological Footprints and Appropriated Carrying Capacity: What Urban Economics Leaves Out," *Environment and Urbanization* 4, no. 2 (1992): 121–130.

8 Fiona McKenzie and Shelby Canterford, "Demographics for Fire Risk Analysis Regional Victoria and Peri-Urban Melbourne" (Canberra: Department of Environment, Land, Water, and Planning, 2016); Kyle M. Stetler, Tyron J. Venn, and David E. Calkin, "The Effects of Wildfire and Environmental Amenities on Property Values in Northwest Montana, USA," *Ecological Economics* 69, no. 11 (2010): 2233–2243.

9 Ted Steinberg, *Acts of God: The Unnatural History of Natural Disaster in America: The Unnatural History of Natural Disaster in America* (New York: Oxford University Press, 2006).

10 Greg Bankoff, Uwe Lübken, and Jordan Sand, eds., *Flammable Cities: Urban Conflagration and the Making of the Modern World* (Madison: University of Wisconsin Press, 2012), 10–11.

11 Mell et al., "The Wildland–Urban Interface Fire Problem"; Howard Botts, Thomas Jeffery, Sheila McCabe, Bryan Stueck, and Logan Suhr, "Wildfire Hazard Risk Report: Residential Wildfire Exposure Estimates for the Western United States," CoreLogic, 2015; also see A. M. Gill and S. L. Stephens, "Scientific and Social Challenges for the Management of Fire-Prone Wildland–Urban Interfaces," *Environmental Research Letters* 4 (2009): 034014; Michael L. Mann, Peter Berck, Max A. Moritz, Enric Batllori, James G. Baldwin, Conor K. Gately, and D. Richard Cameron, "Modeling Residential Development in California from 2000 to 2050," *Land Use Policy* 41 (2014): 438–452; V. C. Radeloff, R. B. Hammer, S. I.

Stewart, J. S. Fried, S. S. Holcomb, and J. F. McKeefry, "The Wildland–Urban Interface in the United States," *Ecological Applications* 15, no. 3 (2005): 799–805.
12 McKenzie and Canterford, "Demographics for Fire Risk Analysis."
13 This is one of many fires in South Africa. For a discussion of the pyrogeography and problems in urban planning see: Sally Archibald, David Roy, Brian van Wilgen, and Robert J. Scholes, "What Limits Fire? An Examination of Drivers of Burnt Area in Southern Africa," *Global Change Biology* 15, no. 3 (2009): 613–630; Twigg et al., "Improved Methods for Fire Risk Assessment"; M. J. Murray, "Fire and Ice: Unnatural Disasters and the Disposable Urban Poor in Post-Apartheid Johannesburg," *International Journal of Urban and Regional Research* 33, no. 1 (2009): 165–192; Sheldon Strydom and Michael J. Savage, "A Spatio-Temporal Analysis of Fires in South Africa," *South African Journal of Science* 112, no. 11–12 (2016): 1–8; Richard Walls, Gerhard Olivier, and Rodney Eksteen, "Informal Settlement Fires in South Africa: Fire Engineering Overview and Full-Scale Tests on 'Shacks,'" *Fire Safety Journal* 91 (2017): 997–1006.
14 Pedro Reszka and Andrés Fuentes, "The Great Valparaiso Fire and Fire Safety Management in Chile," *Fire Technology* 51, no. 4 (2015): 753–58. For a review of Chile's problems with wildfire see Gloria Montenegro, Rosanna Ginocchio, Alejandro Segura, John E. Keely, and Miguel Gómez, "Fire Regimes and Vegetation Responses in Two Mediterranean-Climate Regions," *Revista Chilena de Historia Natural* 77, no. 3 (2004): 455–464; Xavier Úbeda and Pablo Sarricolea, "Wildfires in Chile: A Review," *Global and Planetary Change* 146 (2016): 152–161.
15 Gill and Stephens, "Scientific and Social Challenges"; Sarah M. McCaffrey and Alan Rhodes, "Public Response to Wildfire: Is the Australian 'Stay and Defend or Leave Early' Approach an Option for Wildfire Management in the United States?," *Journal of Forestry* 107, no. 1 (2009): 9–15.
16 Travis Paveglio, Tony Prato, Douglas Dalenberg, and Tyron Venn, "Understanding Evacuation Preferences and Wildfire Mitigations among Northwest Montana Residents," *International Journal of Wildland Fire* 23, no. 3 (2014): 435–444. See also Stetler, Venn, and Calkin, "The Effects of Wildfire and Environmental Amenities on Property Values."
17 S. E. H. Eriksen and H. K. Watson, "The Dynamic Context of Southern African Savannas: Investigating Emerging Threats and Opportunities to Sustainability," *Environmental Science & Policy* 12, no. 1 (2009): 5–22; Pradosh K. Nath and Bhagirath Behera, "A Critical Review of Impact of and Adaptation to Climate Change in Developed and Developing Economies," *Environment, Development and Sustainability* 13, no. 1 (2011): 141–162; Balgis Osman-Elasha, Emmanuel Chidumayo, and Paul Donfack, "Socio-Economic and Gender Related Aspects of Climate Change in Africa," in *Climate Change and African Forest and Wildfire Resources*, ed. Emmanuel Chidumayo, David Okali, Godwin Kowero, and Mahamane Larwanou (Nairobi, Kenya: African Forest Forum, 2011), 176–191, http://www.worldagroforestry.org/downloads/publications/PDFS/BC11263.PDF#page=176.
18 Mark Pelling, ed., *Natural Disaster and Development in a Globalizing World* (London: Routledge, 2003).
19 Hans G. Bohle, Thomas E. Downing, and Michael J. Watts, "Climate Change and Social Vulnerability: Toward a Sociology and Geography of Food Insecurity," *Global Environmental Change* 4, no. 1 (1994): 37–48.

20 Jonathan Crush, "Linking Food Security, Migration and Development," *International Migration* 51, no. 5 (2013): 61–75; Josep Piñol, Jaume Terradas, and Francisco Lloret, "Climate Warming, Wildfire Hazard, and Wildfire Occurrence in Coastal Eastern Spain," *Climatic Change* 38, no. 3 (1998): 345–357; Sirio Modugno, Heiko Balzter, Beth Cole, and Pasquale Borrelli, "Mapping Regional Patterns of Large Forest Fires in Wildland–Urban Interface Areas in Europe," *Journal of Environmental Management* 172 (2016): 112–126.
21 Stephen J. Pyne, *Fire: A Brief History* (Seattle: University of Washington Press, 2001), 115.
22 Nikolas C. Heynen, Maria Kaika, and Erik Swyngedouw, eds., *In the Nature of Cities: Urban Political Ecology and the Politics of Urban Metabolism* (New York: Routledge, 2006).
23 Greg Bankoff, "Introduction," in *Flammable Cities: Urban Conflagration and the Making of the Modern World*, ed. Greg Bankoff, Uwe Lübken, and Jordan Sand (Madison: University of Wisconsin Press, 2012), 3–20.
24 Lewis Mumford, *The City in History* (San Diego, CA: Harcourt, 1961), 283; Ulf Christian Ewert, "Water, Public Hygiene and Fire Control in Medieval Towns: Facing Collective Goods Problems while Ensuring the Quality of Life," *Historical Social Research/Historische Sozialforschung* 32, no. 4 (122) (2007): 222–251.
25 Daniel E. Turbeville, "Cities of Kindling: Geographical Implications of the Urban Fire Hazard on the Pacific Northwest Coast Frontier, 1851–1920" (PhD thesis, Simon Fraser University, 1985).
26 Carl Smith, *The Plan of Chicago: Daniel Burnham and the Remaking of the American City* (Chicago: University of Chicago Press, 2009), 12; Henry Leffmann, "Conflagrations in Cities," in *Proceedings of the Engineers' Club of Philadelphia* 21 (1904): 179–192.
27 Christian Pfister, "Learning from Nature-Induced Disasters," in *Natural Disasters, Cultural Responses: Case Studies toward a Global Environmental History*, ed. Christof Mauch and Christian Pfister (Lanham, MD: Lexington Books, 2009), 17–40.
28 Henri Pirenne, *Medieval Cities: Their Origins and the Revival of Trade*, trans. Frank D. Halsey, vol. 152 (Princeton, NJ: Princeton University Press, 1925); Stefan Krätke, "Cities in Contemporary Capitalism," *International Journal of Urban and Regional Research* 38, no. 5 (2014): 1660–1677.
29 Plutarch, *Plutarch's Lives: The Translation Called Dryden's Corrected from the Greek* (New York: Cosimo, 2013), 205.
30 Currently 70 percent of firefighters in the United States are volunteers. Elsewhere in the world, such as Australia, Chile, and Germany, there is still substantial reliance of volunteer firefighters. In some cases, the use of volunteers was a form of modernization. Historian Samuel Martland has detailed the way in which the Chilean government sought to tame fire in Valparaíso in the nineteenth century. He notes that urban growth and industrialization forced the government to "modernize" its firefighting force. See Samuel Martland, "Taming Fire in Valparaiso, Chile, 1840s–1870s," in *Flammable Cities: Urban Conflagration and the Making of the Modern World*, ed. Greg Bankoff, Uwe Lübken, and Jordan Sand (Madison: University of Wisconsin Press, 2012), 63–81.
31 Amy S. Greenberg, *Cause for Alarm: The Volunteer Fire Department in the Nineteenth-Century City* (Princeton, NJ: Princeton University Press, 1998), 12.

32 Joe Flood, *The Fires: How a Computer Formula, Big Ideas, and the Best of Intentions Burned Down New York City—and Determined the Future of Cities* (New York: Penguin, 2010); Mike Davis, *Ecology of Fear: Los Angeles and the Imagination of Disaster* (New York: Metropolitan Books, 1998).

33 Bernard E. Leete, *Forest Fires in Ohio: 1923 to 1935* (Wooster, OH: Ohio Agricultural Experiment Station, 1939).

34 Timothy Egan, *The Big Burn: Teddy Roosevelt and the Fire That Saved America* (New York: Houghton Mifflin Harcourt, 2009).

35 Christian A. Kull, *Isle of Fire: The Political Ecology of Landscape Burning in Madagascar* (Chicago: University of Chicago Press, 2004).

36 Daniel T. Lichter and David L. Brown, "Rural America in an Urban Society: Changing Spatial and Social Boundaries," *Annual Review of Sociology* 37 (2011): 565–592.

37 Radeloff et al., "The Wildland–Urban Interface in the United States"; David M. Theobald and William H. Romme, "Expansion of the U.S. Wildland–Urban Interface," *Landscape and Urban Planning* 83, no. 4 (2007): 340–354.

38 A. G. Rebelo, P. M. Holmes, C. Dorse, and J. Wood, "Impacts of Urbanization in a Biodiversity Hotspot: Conservation Challenges in Metropolitan Cape Town," *South African Journal of Botany* 77, no. 1 (2011): 20–35; Brian W. van Wilgen, "Fire Management in Species-Rich Cape Fynbos Shrublands," *Frontiers in Ecology and the Environment* 11, no. s1 (2013): e35–44.

39 Joann Gonchar, "Continuing Education: Wildfire-Adapted Design," *Architectural Record*, April 1, 2018, https://www.architecturalrecord.com/articles/13327-continuing-education-wildfire-adapted-design; Chris Woodyard, "Fireproof Houses: Next Step after California Wildfires Sweep the West," *USA Today*, August 17, 2018, https://www.usatoday.com/story/news/2018/08/17/fireproof-houses-california-wildfires-west/963028002/; Fred A. Bernstein, "California Architects Begin to Assess the Catastrophic Wildfire Damage—and Look to the Future," *Architectural Digest*, October 17, 2017, https://www.architecturaldigest.com/story/california-architects-on-wildfires.

40 Albert S. Fu, "The Façade of Safety in California's Shelter-in-Place Homes: History, Wildfire and Social Consequence," *Critical Sociology* 39, no. 6 (2013): 833–849; Pierre Mukheibir and Gina Ziervogel, "Developing a Municipal Adaptation Plan (MAP) for Climate Change: The City of Cape Town," *Environment and Urbanization* 19, no. 1 (2007): 143–158; T. B. Paveglio, M. S. Carroll, and P. J. Jakes, "Alternatives to Evacuation during Wildland Fire: Exploring Adaptive Capacity in One Idaho Community," *Environmental Hazards* 9, no. 4 (2010): 379–394; Genevieve Q. K. Pence, Mark A. Botha, and Jane K. Turpie, "Evaluating Combinations of On-and Off-Reserve Conservation Strategies for the Agulhas Plain, South Africa: A Financial Perspective," *Biological Conservation* 112, no. 1–2 (2003): 253–273; S. J. Rahlao, S. J. Milton, K. J. Esler, B. W. van Wilgen, and P. Barnard, "Effects of Invasion of Fire-Free Arid Shrublands by a Fire-Promoting Invasive Alien Grass (*Pennisetum setaceum*) in South Africa," *Austral Ecology* 34, no. 8 (2009): 920–928.

41 A. Z. Linares and H. A. Linares, "Burn Prevention: The Need for a Comprehensive Approach," *Burns* 16, no. 4 (August 1990): 281–285; Y. G. Şahin and T. İnce, "Early Forest Fire Detection Using Radio-Acoustic Sounding System," *Sensors* 9, no. 3 (2009): 1485–1498; J. H. Sorensen, B. L. Shumpert, and B. M. Vogt, "Planning for Protective Action Decision Making: Evacuate or Shelter-in-Place,"

Journal of Hazardous Materials 109, no. 1–3 (2004): 1–11; Scott L. Stephens, Mark A. Adams, John Handmer, Faith R. Kearns, Bob Leicester, Justin Leonard, and Max A. Moritz, "Urban–Wildland Fires: How California and Other Regions of the U.S. Can Learn from Australia," *Environmental Research Letters* 4 (2009): 1–5; Mukheibir and Ziervogel, "Developing a Municipal Adaptation Plan (MAP) for Climate Change"; Rahlao et al., "Effects of Invasion of Fire-Free Arid Shrublands."

42 Fu, "The Façade of Safety."

43 Leif Gustavsson and Roger Sathre, "Variability in Energy and Carbon Dioxide Balances of Wood and Concrete Building Materials," *Building and Environment* 41, no. 7 (2006): 940–951.

44 Wildfire Defense Systems, homepage, accessed August 30, 2018, https://wildfire-defense.com/.

45 In November 2018, Kanye West and Kim Kardashian brought heighted attention to private firefighters after hiring them to protect their home in Calabasas, California. Karen Bradshaw and Dean Lueck, *Wildfire Policy: Law and Economics Perspectives* (New York: Routledge, 2012); McKenzie Funk, "Too Big to Burn: AIG Plays God in a Man-Made Firestorm," *Harper's Magazine*, October 2009; M. P. McQueen, "Insurers Cater to High-End Homeowners; Pricey Services Are Offered for Special Protection Even before Hurricanes, Wildfires," *Wall Street Journal*, July 1, 2006; Leslie Scism, "Insurers Enlist Private Firefighters," *Wall Street Journal*, November 6, 2017, Eastern edition.

46 Not unlike other forms of insurance, catastrophe insurance coverage varies significantly by country. For instance, France, Spain, and Switzerland have national property insurance programs.

47 *The Proceedings of the Annual Meeting of the Building Officials' Conference,* Pittsburgh, Pennsylvania, February 6–9, 1919; G. Gordon Whitnall, "The Significance of the Recent Supreme Court Zoning Ruling," *Los Angeles Realtor*, April 1925; "Notes and Comment," *Southwest Builder & Contractor*, April 20, 1923.

48 Kathleen Pender, "New State Tax on Wood Products Confusing," *SFGate*, January 10, 2013, https://www.sfgate.com/business/networth/article/New-state-tax-on-wood-products-confusing-4181375.php; George Skelton, "California Lawmakers Are Finally Starting to See the Light through the Soot," *Los Angeles Times*, August 27, 2018, http://www.latimes.com/politics/la-pol-ca-skelton-wildfire-legislation-20180827-story.html.

49 McCaffrey and Rhodes, "Public Response to Wildfire"; Ashton B. Carter, Michael M. May, and William J. Perry, "The Day After: Action following a Nuclear Blast in a U.S. City," *Washington Quarterly* 30, no. 4 (2007): 19–32; M. Ward, J. A. Siegel, and R. L. Corsi, "The Effectiveness of Stand Alone Air Cleaners for Shelter-in-Place," *Indoor Air* 15, no. 2 (2005): 127–134; Thomas J. Cova, "Public Safety in the Urban–Wildland Interface: Should Fire-Prone Communities Have a Maximum Occupancy?," *Natural Hazards Review* 6, no. 3 (2005): 99–108; Thomas J. Cova, Frank A. Drews, Laura K. Siebeneck, and Adrian Musters, "Protective Actions in Wildfires: Evacuate or Shelter-in-Place?," *Natural Hazards Review* 10 (2009): 151.

50 Tom Griffiths, *Forests of Ash: An Environmental History* (Cambridge: Cambridge University Press, 2001); Cova et al., "Protective Actions in Wildfires"; McCaffrey and Rhodes, "Public Response to Wildfire"; Stephens et al., "Urban–Wildland Fires."

51 Craig Gustafson, "Fire Tax's Defeat Puts Supporters in a Bind," *San Diego Union-Tribune*, November 6, 2008, https://www.rsf-fire.org/wp-content/uploads/2017/04/1108-Articles.pdf; Jacqueline Vaughn and Hanna J. Cortner, "Funding Fire: A Losing Proposition?," *California Journal of Politics and Policy* 2, no. 1 (2010); Chris Nichols, "County Continues Backcountry Fire Consolidation," *North County Times*, January 25, 2011; CalFire, "State Responsibility Area Fire Prevention Fee," accessed August 18, 2021, https://www.fire.ca.gov/grants/fire-prevention-grants/state-responsibility-area-fire-prevention-fee/.
52 Riverside County Fire Department, "Conservation Camps," accessed August 8, 2012, http://www.rvcfire.org/opencms/facilities/Camps/.
53 Emily Cadei, "California's Wildfire Risk Is Rising. Congress Missed a Chance to Help," *Sacramento Bee*, February 7, 2018, https://www.sacbee.com/news/politics-government/capitol-alert/article198962219.html.
54 In addition to California, Oregon has created such funds. See Kristen Orwig, "Strategies for Funding Wildfire Mitigation," *Risk Management* 63, no. 5 (June 2016): 14–15.
55 T. B. Paveglio, M. S. Carroll, and P. J. Jakes, "Adoption and Perceptions of Shelter-in-Place in California's Rancho Santa Fe Fire Protection District," *International Journal of Wildland Fire* 19, no. 6 (2010): 677–688.
56 Stetler, Venn, and Calkin, "The Effects of Wildfire and Environmental Amenities on Property Values"; Balaban and Fu, "Politics of Urban Development and Wildfires."
57 CalFire, "Top 20 Largest California Wildfire," accessed January 12, 2018, http://www.fire.ca.gov/communications/downloads/fact_sheets/Top20_Acres.pdf.
58 Jesse Abrams, Katherine Wollstein, and Emily Jane Davis, "State Lines, Fire Lines, and Lines of Authority: Rangeland Fire Management and Bottom-up Cooperative Federalism," *Land Use Policy* 75 (2018): 252–259.
59 That said, coordination is an important component of fighting fire. While the Australian fire defense does not suffer the same neoliberal abdication of responsibility, there are conflicting views in Australia regarding the appropriate role of public agencies, communities, and citizens in dealing with wildfire. See Blythe McLennan and Michael Eburn, "Exposing Hidden-Value Trade-Offs: Sharing Wildfire Management Responsibility between Government and Citizens," *International Journal of Wildland Fire* 24, no. 2 (2015): 162–169; There is a literature regarding the conflict between the mission of Australian agencies. Notably, firefighting goals may conflict with conservation efforts. Also see Å Boholm, Hervé Corvellec, and Marianne Karlsson, "The Practice of Risk Governance: Lessons from the Field," *Journal of Risk Research* 15, no. 1 (2012): 1–20; E. Moskwa, D. Bardsley, D. Weber, and G. Robinson, "Living with Bushfire: Recognising Ecological Sophistication to Manage Risk While Retaining Biodiversity Values," *International Journal of Disaster Risk Reduction* 27 (2018): 459–69.
60 Sorensen, Shumpert, and Vogt, "Planning for Protective Action Decision Making."
61 C. Eriksen, "Why Do They Burn the 'Bush'? Fire, Rural Livelihoods, and Conservation in Zambia," *The Geographical Journal* 173, no. 3 (2007): 242–256; Kull, *Isle of Fire*; A. Shlisky, J. Waugh, P. Gonzalez, M. Gonzalez, M. Manta, H. Santoso, E. Alvarado et al., "Fire, Ecosystems and People: Threats and Strategies for Global Biodiversity Conservation," *The Nature Conservancy*, 2007; T. Collins,

"The Political Ecology of Hazard Vulnerability," *Journal of Political Ecology* 15, no. 1 (2008): 21–43.

62 Megan Cassidy and Melody Gutierrez, "As California Burns, Volunteer Firefighters Become Harder to Find," *San Francisco Chronicle*, August 11, 2018, https://www.sfchronicle.com/california-wildfires/article/As-California-burns-volunteer-firefighters-13148559.php; Angie Sharp, "Digging Deeper into the Volunteer Firefighter Shortage," *WQAD.com*, April 26, 2018, https://wqad.com/2018/04/26/digging-deeper-into-the-volunteer-firefighter-shortage/.

63 Loïc J. D. Wacquant, *Prisons of Poverty* (Minneapolis: University of Minnesota Press, 2009).

64 Anna Doty, "Fireline, Divided: Labor Representation of Unionized and Incarcerated Firefighters in California's Wildlands" (master's thesis, Massachusetts Institute of Technology, 2017); Philip Goodman, "'Another Second Chance' Rethinking Rehabilitation through the Lens of California's Prison Fire Camps," *Social Problems* 59, no. 4 (2012): 437–458; also see Tracy Connor, "Firefighting Felons: Hundreds of Inmates Battling the Yosemite Blaze," *NBCNews.com*, August 30, 2014, http://www.nbcnews.com/news/other/firefighting-felons-hundreds-inmates-battling-yosemite-blaze-f8C11039218; Alex Helmick, "California Leans Heavily on Thousands of Inmate Firefighters," *KQED.org*, July 25, 2014, http://blogs.kqed.org/science/2014/07/25/california-leans-heavily-on-thousands-of-inmate-firefighters/; Clark Mindock, "As California's Biggest Ever Wildfire Burns, Should Inmates Be Fighting the Blaze?," *The Independent*, August 8, 2018, https://www.independent.co.uk/news/world/americas/california-wildfires-latest-inmate-firefighters-prisoners-mendocino-complex-carr-fire-a8483571.html.

65 Philip Goodman, "Hero and Inmate: Work, Prisons, and Punishment in California's Fire Camps," *WorkingUSA* 15, no. 3 (2012): 353–376; also see Foon Rhee, "A Candidate for Lieutenant Governor Says Inmate Firefighters Are Slave Labor. Is She Right?," *Sacramento Bee*, September 20, 2017, http://www.sacbee.com/opinion/opn-columns-blogs/foon-rhee/article174370641.html; Alene Tchekmedyian, "Inmate Firefighter Collapses, Dies on Training Hike during First Day on Job," *Los Angeles Times*, April 23, 2018, http://www.latimes.com/local/lanow/la-me-ln-inmate-firefighter-dies-northern-california-20180423-story.html; Chantal da Silva, "U.S. Inmates Launching Nationwide Strike to 'End Prison Slavery,'" *Newsweek*, August 21, 2018, https://www.newsweek.com/us-inmates-launching-nationwide-prison-strike-end-prison-slavery-1082034; Jaime Lowe, "The Incarcerated Women Who Fight California's Wildfires," *New York Times*, August 31, 2017, Magazine, https://www.nytimes.com/2017/08/31/magazine/the-incarcerated-women-who-fight-californias-wildfires.html; Kamala Kelkar, "Incarcerated Women Risk Their Lives Fighting California Fires. It's Part of a Long History of Prison Labor," *PBS NewsHour*, October 22, 2017, https://www.pbs.org/newshour/nation/incarcerated-women-risk-their-lives-fighting-california-fires-its-part-of-a-long-history-of-prison-labor.

66 National Wildfire Suppression Association, homepage, accessed August 30, 2018, http://www.nwsa.us; David Tao, "The Big Business of Battling Wildfires," *Fortune*, June 22, 2011, http://fortune.com/2011/06/22/the-big-business-of-battling-wildfires/.

67 Ted Goldberg, "A 1960s Law Blocks Firefighting Contractors from Suing State," *KQED.org*, November 8, 2017, https://www.kqed.org/news/11628474/a-1960s-law-blocks-firefighting-contractors-from-suing-state.

68 Steve Duin, "Forestry Contracting and Prevailing Greed in Oregon's Woods," *OregonLive.com*, October 31, 2011, http://www.oregonlive.com/news/oregonian/steve_duin/index.ssf/2011/10/federal_contracting_and_prevai.html; Josh McDaniel and Vanessa Casanova, "Forest Management and the H2B Guest Worker Program in the Southeastern United States: An Assessment of Contractors and Their Crews," *Journal of Forestry* 103, no. 3 (2005): 114–19.

69 Brinda Sarathy, *Pineros: Latino Labour and the Changing Face of Forestry in the Pacific Northwest* (Vancouver: UBC Press, 2012), 10–11. Also see U.S. Department of Homeland Security and U.S. Citizenship and Immigration Services, "Characteristics of H-2B Nonagricultural Temporary Workers: Fiscal Year 2016 Report to Congress," July 14, 2017, https://cis.org/sites/default/files/2018-03/h2b_annual_report_2016.pdf.

70 Mack Lamoureux, "Singing South African Firefighters Going Home amid Pay Dispute over Fort McMurray Wildfire," *CBC News*, June 12, 2016, https://www.cbc.ca/news/canada/edmonton/south-african-firefighters-in-pay-dispute-leave-fort-mcmurray-1.3631385.

71 Dim Coumou and Stefan Rahmstorf, "A Decade of Weather Extremes," *Nature Climate Change* 2, no. 7 (2012): 491–496; Westerling and Bryant, "Climate Change and Wildfire in California."

72 Nancy B. Grimm, Stanley H. Faeth, Nancy E. Golubiewski, Charles L. Redman, Jianguo Wu, Xuemei Bai, and John M. Briggs, "Global Change and the Ecology of Cities," *Science* 319, no. 5864 (2008): 756–760; D. E. Pataki, R. J. Alig, A. S. Fung, N. E. Golubiewski, C. A. Kennedy, E. G. McPherson, D. J. Nowak et al., "Urban Ecosystems and the North American Carbon Cycle," *Global Change Biology* 12, no. 11 (2006): 2092–2102.

73 Steve Fraser and Joshua B. Freeman, "Fires, Floods, Earthquakes, and Capitalism," *New Labor Forum* 22, no. 2 (2013): 96–99.

Chapter 5   Assessing and Managing Risk

1 Stephanie Farmer, "Uneven Public Transportation Development in Neoliberalizing Chicago, USA," *Environment and Planning A* 43, no. 5 (2011): 1154–1172; Josh Pacewicz, "Tax Increment Financing, Economic Development Professionals and the Financialization of Urban Politics," *Socio-Economic Review* 11, no. 3 (2012): 413–440; Kevin F. Gotham, "The Secondary Circuit of Capital Reconsidered: Globalization and the U.S. Real Estate Sector," *American Journal of Sociology* 112, no. 1 (2006): 231–275.

2 Sarah Knuth, "Green Devaluation: Disruption, Divestment, and Decommodification for a Green Economy," *Capitalism Nature Socialism* 28, no. 1 (2017): 98–117; Sarah Elisabeth Knuth, "Global Finance and the Land Grab: Mapping Twenty-First Century Strategies," *Canadian Journal of Development Studies/Revue Canadienne d'études Du Développement* 36, no. 2 (2015): 163–178.

3 Desiree Fields, "Urban Struggles with Financialization," *Geography Compass* 11, no. 11 (2017): e12334.

4 "Transforming Our World: The 2030 Agenda for Sustainable Development," United Nations Resolution, October 21, 2015; "Decoupling Natural Resource Use and Environmental Impacts from Economic Growth," United Nations Environment Program, 2011, https://www.resourcepanel.org/reports/decoupling-natural

-resource-use-and-environmental-impacts-economic-growth; OECD, "Indicators to Measure Decoupling of Environmental Pressure from Economic Growth," Ad Hoc Group for Sustainable Development, May 16, 2002.
5. T. Parrique, J. Barth, F. Briens, C. Kerschner, A. Kraus-Polk, A. Kuokkanen, and J. H. Spangenberg, *Decoupling Debunked* (Brussels: European Environmental Bureau, 2019).
6. Andrew K. Jorgenson and Brett Clark, "Are the Economy and the Environment Decoupling? A Comparative International Study, 1960–2005," *American Journal of Sociology* 118, no. 1 (2012): 1–44; Julia K. Steinberger, Fridolin Krausmann, Michael Getzner, Heinz Schandl, and Jim West, "Development and Dematerialization: An International Study," *PLoS One* 8, no. 10 (2013): e70385.
7. Joseph M. Simpson, Riley E. Dunlap, and Andrew S. Fullerton, "The Treadmill of Information: Development of the Information Society and Carbon Dioxide Emissions," *Sociology of Development* 5, no. 4 (2019): 381–409.
8. The Gramm–Leach–Bliley Act, also known as the Financial Services Modernization Act of 1999, removed barriers between banking, brokerage, and insurance companies.
9. Ted Steinberg, "The Secret History of Natural Disaster," *Global Environmental Change Part B: Environmental Hazards* 3, no. 1 (2001): 31–35; Tristan Sturm and Eric Oh, "Natural Disasters as the End of the Insurance Industry? Scalar Competitive Strategies, Alternative Risk Transfers, and the Economic Crisis," *Geoforum* 41, no. 1 (2010): 154–163.
10. Naomi Klein, *This Changes Everything: Capitalism vs. the Climate* (New York: Simon and Schuster, 2014), 234.
11. Gregory D. Squires, "Racial Profiling, Insurance Style: Insurance Redlining and the Uneven Development of Metropolitan Areas," *Journal of Urban Affairs* 25, no. 4 (2003): 391–410.
12. NAIC, "Municipal Bonds," *NAIC.org*, March 4, 2020, https://content.naic.org/cipr_topics/topic_municipal_bonds.htm.
13. Evan Mills, "Synergisms between Climate Change Mitigation and Adaptation: An Insurance Perspective," *Mitigation and Adaptation Strategies for Global Change* 12, no. 5 (2007): 809–842.
14. Jane Milne, "Climate Change, Insurance and the Building Sector: Synergisms, Conflicts and Adaptive Capacity," *Building Research & Information* 32, no. 1 (2004): 48–54.
15. Steve Evans, "AIG Sees $500m California Wildfire Loss, Expects to Buy More Reinsurance," *Seeking Alpha*, November 7, 2017, https://seekingalpha.com/article/4121946-aig-sees-500m-california-wildfire-loss-expects-buy-reinsurance. As described by Karl Marx, "It matters not whether this insurance fund is managed by insurance companies as a separate business or not. This is the sole portion of revenue which is neither consumed as such nor serves necessarily as a fund for accumulation. Whether it actually serves as such, or covers merely a loss in reproduction, depends upon chance." See Karl Marx, *Capital: Volume III* (Moscow: International Publishers, 1959), 608, https://www.marxists.org/archive/marx/works/1894-c3/index.htm.
16. Howard Kunreuther and Erwann O. Michel-Kerjan, *At War with the Weather* (Boston: MIT Press, 2009).
17. Michael Lewis, "In Nature's Casino," *New York Times*, August 26, 2007, http://www.nytimes.com/2007/08/26/magazine/26neworleans-t.html.

18 The Chicago Mercantile Exchange Group is a product of a 2007 merger between the Chicago Mercantile Exchange and the Chicago Board of Trade (CBOT). CBOT, established in 1848, is one of the world's oldest futures exchanges.
19 Kevin Grove, "Preempting the Next Disaster: Catastrophe Insurance and the Financialization of Disaster Management," *Security Dialogue* 43, no. 2 (2012): 139–155.
20 Stefan Hochrainer and Reinhard Mechler, "Natural Disaster Risk in Asian Megacities: A Case for Risk Pooling?," *Cities* 28, no. 1 (2011): 53–61.
21 SwissRe Institute, "Global Economic and Insurance Outlook 2020," *Sigma*, May 2018, 11.
22 Kristen Orwig, "Strategies for Funding Wildfire Mitigation," *Risk Management* 63, no. 5 (June 2016): 14–15; Jeb Brugmann, "Financing the Resilient City," *Environment and Urbanization* 24, no. 1 (2012): 215–232.
23 Shalini Vajhala and James Rhodes, "Leveraging Catastrophe Bonds: As a Mechanism for Resilience Infrastructure Project Finance," Refocus Partners, February 2017, 12, http://www.refocuspartners.com/wp-content/uploads/2017/02/RE.bound-Program-Report-December-2015.pdf; OECD, *Investing in Climate, Investing in Growth* (Paris: OECD Publishing, 2017), 291–292.
24 Vajhala and Rhodes, "Leveraging Catastrophe Bonds," 12; OECD, *Investing in Climate, Investing in Growth*, 291–292.
25 Heike Reichelt and Colleen Keenan, "The Green Bond Market: 10 Years Later and Looking Ahead," World Bank, December 2017, 27.
26 European Commission, "Action Plan: Financing Sustainable Growth," March 2018, https://eur-lex.europa.eu/legal-content/EN/TXT/PDF/?uri=CELEX:52018DC0097&from=EN.
27 Aon Benfield, "Reinsurance Market Outlook: Reinsurance Proves Its Worth," Aon plc, January 2018, 5–8, http://thoughtleadership.aonbenfield.com/Documents/20180103-ab-analytics-rmo-january.pdf.
28 Dick Bryan and Michael Rafferty, "Deriving Capital's (and Labour's) Future," in *The Crisis This Time: Socialist Register 2011*, ed. Greg Albo, Leo Panitch, and Vivek Chibber (London: Monthly Review Press, 2010), 198.
29 EIOPA, *Financial Stability Report Second Half-Year Report: Autumn 2013* (Frankfurt: European Insurance and Occupational Pensions Authority, December 2013), 16, https://www.eiopa.europa.eu/sites/default/files/publications/reports/fsr-december2013.pdf.
30 Paul J. Davies, "Hurricane Irma Tests Risk-Reward of Catastrophe-Bond Market," *Wall Street Journal*, September 8, 2017, Markets, https://www.wsj.com/articles/hurricane-irma-tests-risk-reward-of-catastrophe-bond-market-1504895037.
31 Oliver Ralph, "Global Catastrophe Bond Market Size Climbs to a Record $30bn," *Financial Times*, September 7, 2018, https://www.ft.com/content/d62827b2-b1e0-11e8-99ca-68cf89602132; also see Mengqi Sun and Leslie Scism, "Even after Last Year's Terrible Hurricanes, Insurers Are in Solid Shape," *Wall Street Journal*, June 30, 2018, Markets, https://www.wsj.com/articles/even-after-last-years-terrible-hurricanes-insurers-are-in-solid-shape-1530365549; Artemis, "Oppenheimer Cat Bond Fund AuM up, as 2017 Losses Still Being Realised," *Artemis.Bm* (blog), June 11, 2018, http://www.artemis.bm/blog/2018/06/11/oppenheimer-cat-bond-fund-aum-up-as-2017-losses-still-being-realised/; OppenheimerFunds, "Catastrophe (CAT) Bond Strategy," accessed October 25, 2018, https://www.oppenheimerfunds.com/private-client-groups/investment-strategies/cat-bond/ELBS.

32 EIOPA, *Financial Stability Report: Key Developments* (Frankfurt: European Insurance and Occupational Pensions Authority, June 2018), 8, https://www.eiopa.europa.eu/content/financial-stability-report-june-2018_en.
33 SwissRe Institute, "Global Economic and Insurance Outlook 2020."
34 Adriana X. Sanchez, Jeroen van der Heijden, and Paul Osmond, "The City Politics of an Urban Age: Urban Resilience Conceptualisations and Policies," *Palgrave Communications* 4, no. 1 (2018): 25.
35 Sarah Dooling and Gregory Simon, eds., *Cities, Nature and Development: The Politics and Production of Urban Vulnerabilities* (Surrey, UK: Ashgate Publishing, 2012); B. E Goldstein, "Skunkworks in the Embers of the Cedar Fire: Enhancing Resilience in the Aftermath of Disaster," *Human Ecology* 36, no. 1 (2008): 15–28; Mark Pelling, *The Vulnerability of Cities: Natural Disasters and Social Resilience* (London: Earthscan, 2003).
36 Scott Frickel and James R. Elliott, *Sites Unseen: Uncovering Hidden Hazards in American Cities* (New York: Russell Sage Foundation, 2018), 23–24.
37 Derek K. Kellenberg and Ahmed Mushfiq Mobarak, "Does Rising Income Increase or Decrease Damage Risk from Natural Disasters?," *Journal of Urban Economics* 63, no. 3 (2008): 788–802.
38 Sturm and Oh, "Natural Disasters as the End of the Insurance Industry?"; Kathleen J. Tierney, "Toward a Critical Sociology of Risk," *Sociological Forum* 14 (1999): 215–242.
39 Pelling, *The Vulnerability of Cities*, 8.
40 FEMA, "What Is Mitigation?," accessed July 5, 2018, https://www.fema.gov/what-mitigation.
41 Mike Davis, "Who Will Build the Ark?," *New Left Review* 61 (2010): 29–46; L. Hunter Lovins and Boyd Cohen, *Climate Capitalism: Capitalism in the Age of Climate Change* (New York: Farrar, Straus and Giroux, 2011), 95–96.
42 Lovins and Cohen, *Climate Capitalism*, 95–99.
43 Furthermore, Gould and Lewis warn that post–Hurricane Irma, places such as Barbuda could see redevelopment for the global elite. This would undermine the country's self-sufficiency and reconfigure its institutions toward capitalistic growth and consumption. See Kenneth A. Gould and Tammy L. Lewis, *Green Gentrification: Urban Sustainability and the Struggle for Environmental Justice* (New York: Routledge, 2016); Kenneth A. Gould and Tammy L. Lewis, "Green Gentrification and Disaster Capitalism in Barbuda," *NACLA Report on the Americas* 50, no. 2 (2018): 148–153.
44 Jesse Goldstein, "Appropriate Technocracies? Green Capitalist Discourses and Post Capitalist Desires," *Capitalism Nature Socialism* 24, no. 1 (2013): 28.
45 Ashley Dawson, *Extreme Cities: The Peril and Promise of Urban Life in the Age of Climate Change* (London: Verso Books, 2017).
46 Brundtland Commission, "Our Common Future: Report of the World Commission on Environment and Development," U.N. Document A/42/427, chap. 2, http://www.un-documents.net/ocf-02.htm.
47 UNISDR, *How to Make Cities More Resilient* (Geneva: United Nations, 2012), 51.
48 Jack Ahern, "From Fail-Safe to Safe-to-Fail: Sustainability and Resilience in the New Urban World," *Landscape and Urban Planning* 100, no. 4 (2011): 341–343.
49 Simin Davoudi, "Just Resilience," *City & Community* 17, no. 1 (2018): 3–7; Simin Davoudi, "Resilience: A Bridging Concept or a Dead End," *Planning Theory & Practice* 13, no. 2 (2012): 299–307.

50 Melinda Harm Benson and Robin Kundis Craig, *The End of Sustainability: Resilience and the Future of Environmental Governance in the Anthropocene* (Lawrence: University Press of Kansas, 2017).
51 Richard J. T. Klein, Robert J. Nicholls, and Frank Thomalla, "Resilience to Natural Hazards: How Useful Is This Concept?," *Global Environmental Change Part B: Environmental Hazards* 5, no. 1 (2003): 35–45.
52 Steward T. A. Pickett, Mary L. Cadenasso, and J. Morgan Grove, "Resilient Cities: Meaning, Models, and Metaphor for Integrating the Ecological, Socio-Economic, and Planning Realms," *Landscape and Urban Planning* 69, no. 4 (2004): 369–384.
53 Jonatan A. Lassa, "Roles of Non-Government Organizations in Disaster Risk Reduction," *Oxford Research Encyclopedia of Natural Hazard Science*, June 25, 2018, http://naturalhazardscience.oxfordre.com/view/10.1093/acrefore/9780199389407.001.0001/acrefore-9780199389407-e-45.
54 UNIDSR, "Terminology," *UNIDSR.org*, accessed November 14, 2018, https://www.unisdr.org/we/inform/terminology#letter-r.
55 GNDR, "GNDR Factsheet," May 2020, https://www.gndr.org/images/newsite/About_us/GNDR_Factsheet.pdf.
56 International Federation of Red Cross and Red Crescent Societies, "The Road to Resilience: Bridging Relief and Development for a More Sustainable Future," IFRC Discussion Paper, June 2012, 6.
57 GFDRR, "Spring 2017 CG Meeting: City Resilience Program," Zurich, Switzerland, April 5–7, 2017, 3.
58 GFDRR, "City Resilience Program: Annual Report," 2018, 6.
59 Lawrence J. Vale, "The Politics of Resilient Cities: Whose Resilience and Whose City?," *Building Research & Information* 42, no. 2 (2014): 191–201.
60 Timothy Beatley, *Biophilic Cities: Integrating Nature into Urban Design and Planning* (Washington, DC: Island Press, 2011); William McDonough and Michael Braungart, *Cradle to Cradle: Remaking the Way We Make Things* (New York: North Point Press, 2002).
61 Helga Leitner, Eric Sheppard, Sophie Webber, and Emma Colven, "Globalizing Urban Resilience," *Urban Geography* 39, no. 8 (2018): 1277.
62 David Beer, "Productive Measures: Culture and Measurement in the Context of Everyday Neoliberalism," *Big Data & Society* 2, no. 1 (2015): 1–12.
63 GFDRR, *The Making of a Riskier Future: How Our Decisions Are Shaping Future Disaster Risk* (Washington, DC: Global Facility for Disaster Reduction and Recovery, 2016), vii.
64 UNIDSR's handbook also uses ISO 31000. It notes the importance of taking into consideration "human and cultural factors" in managing risk. See UNISDR, *How to Make Cities More Resilient*, 69.
65 Hector E. Valdez and Abdol R. Chini, "ISO 14000 Standards and the U.S. Construction Industry," *Environmental Practice* 4, no. 4 (2002): 210–219.
66 Robert L. Westly and Ralph Vasquez, "LEED and ISO 14001: Working Together," *Building Operating Management*, May 2006, https://docs.google.com/viewerng/viewer?url=https://www.scsengineers.com/wp-content/uploads/2015/03/Westly_Vasquez-LEED_and_ISO_1400_Working_Together.pdf&hl=en_US.
67 Stefan Timmermans and Steven Epstein, "A World of Standards but Not a Standard World: Toward a Sociology of Standards and Standardization," *Annual*

Review of Sociology 36 (2010): 71; see also Leitner et al., "Globalizing Urban Resilience," 1277–1278.
68 Scott Knowles, *The Disaster Experts: Mastering Risk in Modern America* (Philadelphia: University of Pennsylvania Press, 2012), 5.
69 Aidan While, Andrew E. G. Jonas, and David Gibbs, "The Environment and the Entrepreneurial City: Searching for the Urban 'Sustainability Fix' in Manchester and Leeds," *International Journal of Urban and Regional Research* 28, no. 3 (September 2004): 549–569; Steven Lang and Julia Rothenberg, "Neoliberal Urbanism, Public Space, and the Greening of the Growth Machine: New York City's High Line Park," *Environment and Planning A* 49, no. 8 (2017): 1743–1761.
70 UNISDR, *How to Make Cities More Resilient*, 15.
71 The Intergovernmental Panel on Climate Change (IPCC) makes use of the word "synergy" in discussing reducing emissions. See IPCC, "12.3 Implications of Mitigation Choices for Sustainable Development Goals," in *Climate Change 2007: Mitigation of Climate Change* (Cambridge: Cambridge University Press, 2007), 726–729, https://www.ipcc.ch/publications_and_data/ar4/wg3/en/ch12s12-3.html.
72 Rick Gould, "The Secret to Unlocking Green Finance," *ISO.org*, May 8, 2018, http://www.iso.org/cms/render/live/en/sites/isoorg/contents/news/2018/05/Ref2287.html.
73 Benjamin Cashore, "Legitimacy and the Privatization of Environmental Governance: How Non-State Market-Driven (NSMD) Governance Systems Gain Rule-Making Authority," *Governance* 15, no. 4 (2002): 503–529; Jennifer Clapp, "The Privatization of Global Environmental Governance: ISO 14000 and the Developing World," *Global Governance* 4, no. 3 (1998): 295–316; also see Ryan Gunderson, "Global Environmental Governance Should Be Participatory: Five Problems of Scale," *International Sociology* 33, no. 6 (2018): 715–737.
74 Douglas Pierce, *RELi Resilience Action List and Credit Catalog v1.2.1* (C3 Living Design + Captain Markets Partnership, June 5, 2017), https://s3.amazonaws.com/online.anyflip.com/zyqc/ojoi/mobile/index.html; G. Abdalla, G. J. Maas, J. Huyghe, and M. Oostra, "Criticism on Environmental Assessment Tools," in *Proceeding of 2nd International Conference on Environmental Science and Technology* (Singapore: Institute of Electrical and Electronics Engineers, 2011), 4, https://research.tue.nl/en/publications/criticism-on-environmental-assessment-tools.
75 Guy R. Newsham, Sandra Mancini, and Benjamin J. Birt, "Do LEED-Certified Buildings Save Energy? Yes, But . . . ," *Energy and Buildings* 41, no. 8 (2009): 897–905.
76 Richard York, "Ecological Paradoxes: William Stanley Jevons and the Paperless Office," *Human Ecology Review* 13, no. 2 (2006): 143–147.

Conclusion

1 T. Mitchell, *Carbon Democracy: Political Power in the Age of Oil* (London: Verso Books, 2011).
2 Herbert Girardet, *Creating Regenerative Cities* (London: Routledge, 2014).
3 James Wesley Scott, "Smart Growth as Urban Reform: A Pragmatic 'Recoding' of the New Regionalism," *Urban Studies* 44, no. 1 (2007): 15–35.

4   Robert Pollin argues that de-growth is not a viable strategy. Looking at market contractions during the Great Recession (2007–2009), he suggests there would need to be a reduction of gross national product (GDP) four times greater than occurred during the financial crisis to reduce emissions in line with the recommendations of the Intergovernmental Panel on Climate Change (IPCC). Instead, Pollin and others have called for a green revolution or "Green New Deal." See Robert Pollin, "De-Growth vs. a Green New Deal," *New Left Review* 12 (July/August 2018): 5–25.
5   David Harvey, *Spaces of Hope* (Berkeley: University of California Press, 2000), 35.
6   Henri Lefebvre, *The Urban Revolution* (Minneapolis: University of Minnesota Press, 2003), 90, also 25–27, 104.
7   Girardet, *Creating Regenerative Cities*; Giles Thomson and Peter Newman, "Cities and the Anthropocene: Urban Governance for the New Era of Regenerative Cities," *Urban Studies* 57, no. 7 (2020), 1502–1519; Giles Thomson and Peter Newman, "Urban Fabrics and Urban Metabolism—from Sustainable to Regenerative Cities," *Resources, Conservation and Recycling* 132 (2018): 218–229.
8   Christof Mauch, "Introduction," in *Natural Disasters, Cultural Responses*, ed. Christof Mauch and Christian Pfister (Lanham, MD: Lexington Books, 2009), 1–16.
9   Pia Hollenbach and Kanchana N. Ruwanpura, "Symbolic Gestures: The Development Terrain of Post-Tsunami Villages in (Southern) Sri Lanka," *Journal of Development Studies* 47, no. 9 (2011): 1299–1314; Edward Simpson and Stuart Corbridge, "The Geography of Things That May Become Memories: The 2001 Earthquake in Kachchh-Gujarat and the Politics of Rehabilitation in the Prememorial Era," *Annals of the Association of American Geographers* 96, no. 3 (2006): 566–585.
10 Kathleen Tierney and Anthony Oliver-Smith, "Social Dimensions of Disaster Recovery," *International Journal of Mass Emergencies and Disasters* 30, no. 2 (2012): 125.
11 David Harvey, *Rebel Cities: From the Right to the City to the Urban Revolution* (London: Verso Books, 2012), xvii.
12 "Santa Clara: Housing Planned on Landfill near Levi's Stadium," *Mercury News*, July 8, 2017, https://www.mercurynews.com/2017/07/08/the-latest-silicon-valley-housing-idea-on-a-landfill/.
13 William R. Freudenburg, "Contamination, Corrosion and the Social Order: An Overview," *Current Sociology* 45, no. 3 (1997): 19–39.
14 Henri Lefebvre, *The Production of Space* (Oxford: Basil, 1991), 422.
15 Michael Edema Leary-Owhin, *Exploring the Production of Urban Space: Differential Space in Three Post-Industrial Cities* (Bristol: Policy Press, 2016).
16 Lefebvre, *The Production of Space*, 423, also see 52, and 167.
17 Uta Hassler and Niklaus Kohler, "The Ideal of Resilient Systems and Questions of Continuity," *Building Research & Information* 42, no. 2 (2014): 158–167.
18 Kate Aronoff, Alyssa Battistoni, Daniel Aldana Cohen, and Thea Riofrancos, *A Planet to Win: Why We Need a Green New Deal* (New York: Verso Books, 2019), 127–128.
19 Daniel P. Aldrich, *Building Resilience: Social Capital in Post-Disaster Recovery* (Chicago: University of Chicago Press, 2012), 2–3.
20 Eric Klinenberg, *Palaces for the People: How Social Infrastructure Can Help Fight Inequality, Polarization, and the Decline of Civic Life* (New York: Broadway Books, 2018), 186–187.

21 Christophe Béné, Rachel Godfrey Wood, Andrew Newsham, and Mark Davies, "Resilience: New Utopia or New Tyranny?," *IDS Working Papers* 2012, no. 405 (2012): 1–61.
22 Holly P. Jones, David G. Hole, and Erika S. Zavaleta, "Harnessing Nature to Help People Adapt to Climate Change," *Nature Climate Change* 2, no. 7 (2012): 504.
23 Lefebvre, *The Production of Space*, 167. Also, Lefebvre's description of cities as "oeuvre" is useful here. See Henri Lefebvre, *Writings on Cities*, ed. Elizabeth Lebas (Oxford: Blackwell, 1996), 65–67.
24 Pollin, "De-Growth vs. a Green New Deal."
25 Thomson and Newman have argued that green design, sustainable development, and regenerative urbanism could serve as an alternative form of large-scale geoengineering. See Giles Thomson and Peter Newman, "Geoengineering in the Anthropocene through Regenerative Urbanism," *Geosciences* 6, no. 4 (2016): 46.
26 David Harvey, "The Right to the City," *New Left Review* 53 (September/October 2008): 35.
27 Anna Geddes, Tobias S. Schmidt, and Bjarne Steffen, "The Multiple Roles of State Investment Banks in Low-Carbon Energy Finance: An Analysis of Australia, the UK and Germany," *Energy Policy* 115 (2018): 158–170.
28 Timothy Beatley, *Biophilic Cities: Integrating Nature into Urban Design and Planning* (Washington, DC: Island Press, 2011); William McDonough and Michael Braungart, *Cradle to Cradle: Remaking the Way We Make Things* (New York: North Point Press, 2002).
29 Travis B. Paveglio, Pamela J. Jakes, Matthew S. Carroll, and Daniel R. Williams, "Understanding Social Complexity Within the Wildland–Urban Interface: A New Species of Human Habitation?," *Environmental Management* 43, no. 6 (2009): 1085–1095.
30 Girardet, *Creating Regenerative Cities*, 99. Naomi Klein has talked about the "right to generate." While she is largely talking about reproductive health, she does discuss changing from a model of "extraction" to one of renewal. Naomi Klein, *This Changes Everything: Capitalism vs. the Climate* (New York: Simon and Schuster, 2014), 419.

Index

Aalbers, Manuel B., 10
Actividades de Construcción y Servicios, S.A. (ACS), 65, 69
adaptation, 96, 105, 106
Adorno, Theodor, 95
Advanced Weather Interactive Processing System, 32
Ahern, Jack, 109
AIR Worldwide, 32
Albayrak Group, 68, 69, 70
Alberti, Leon Battista, 42
Aldrich, Daniel, 121
Alford Safety Systems, 92
Allianz company, 86
American International Group (AIG), 86, 98, 99
Anderson Rubbish Disposal, 58
anthropocentric hubris, 55
Arizona: firefighter program in, 91; garbage incineration, 72
Aronoff, Kate, 121
Asian Development Bank (ADB), 102
Australia: Black Summer in, 78, 92; bushfires, 77–78, 79, 80; Chambers Gully Reserve, 120; disaster management in, 88, 154n59; volunteer firefighters in, 151n30

Baker, Richard H., 31
Baltimore Fire (1904), 87

Banham, Reyner, 22
Bankoff, Greg, 82
Barbon, Nicholas, 83
Basel Convention of 1989, 63
Beck, Ulrich, 5
Beer, David, 112
Beijing Enterprise, 74
Béné, Christopher, 122
Benfield, Aon, 3
Benson, Mindy, 110
Biesbroek, G. Robbert, 14
Big Data, 25
Booth, Kate, 24
Boston, MA: Great Fire of 1872, 29
Brundtland Commission, 109
Bryan, Dick, 103
building codes, 87–88, 116
Building Officials Conference (1915), 87
Building Research Establishment Environmental Assessment Method (BREEAM), 6, 112
built environment: capitalism and, 19, 20, 24, 26; energy and water consumption, 107; geographic and spatial limitations to, 29; risk of, 20–21; as technology, 24–25
Bunker, Stephen, 71
Burnham, Daniel, 82
Butt, Waqas H., 13
Buttel, Frederick, 71

California: Assembly Bill 2727, 91; building codes in, 87–88; César Chávez Park, 120; design of homes in, 85; disaster management, 88–89; evacuation orders, 88; Fire Department's budget, 89; firefighter shortages, 91; fire-insurance program, 89–90; fire protection fees, 89; Garden Route, 84; private firefighting companies, 92; vegetation, 84

California Department of Forestry and Fire Protection (CalFire), 89, 91, 93

California Department of Resources Recycling and Recovery (CalRecycle), 75

California State Building Standards Commission, 87

California wildfires: damage from, 89, 90; ecological impact of, 22, 79; as fixture of landscape, 85; mishandling of, 83; recovery from, 79; scale of, 89; social impact of, 36; statistics of, 78; 2018 Camp Fire, 1–2, 8

Cam Pak, 68

Canada: migrant firefighters, 93; wildfires in, 76, 78, 93

Cape Town: Fire and Rescue Services, 84; water crisis, 84

capitalism: as blowback, 35; built environment and, 19, 20, 24, 26, 34; critique of growth-logic of, 120; disaster readiness and, 19, 34; fire risk and, 82; inequality of, 35–36; relationship with nature, 28–29, 74, 108; rhythm of, 18; risk management and, 95–96; urban sustainability and, 4

capitalism-in-nature, 28

Capital Markets Partnership (CMP), 112

carbon capture technologies, 148n94

carbon dioxide (CO_2) emission, 81, 108, 143n30, 162n4

Care Ambulance Service, 92

Caribbean Catastrophe Risk Insurance Facility (CCRIF), 101

Carmin, JoAnn, 27

Carruthers, Bruce G., 24

casino capitalism, 33

catastrophe: average mortality, 2; conditions for, 12; as creative destruction, 30; profit generated by, 35; real estate development and, 30–31; recovery efforts, 36

catastrophe bonds: investment strategies in, 103, 123; market for, 100, 101; types of, 99

catastrophe models: climate data for, 32; commodification of, 32–33; criticism of, 26, 33; dominance of proprietary, 33; insurance policies and, 33; policymaking and, 25–26

Catton, William, 41

Center for Coordination of Natural Disaster Prevention in Central America, 34

Central American Probabilistic Risk Assessment (CAPRA), 34

Ceres Investor Summit on Climate Risk (2018), 10

Chase, Stuart, 58

Chicago, IL: Great Fire of 1871, 5, 82, 83, 94; heat wave of 1995, 99

Chicago Mercantile Exchange (CME), 32, 100

Chile: volunteer firefighters in, 151n30; waste companies, 67; wildfires in, 77, 80

China: Belt and Road Initiative, 68; recycling program, 63, 72; urban sinkholes in, 42–43; waste management companies, 65, 66

Chubb (insurance company), 86

cities: anticipation of potential harm, 21; in Antiquity, 45; in coastal areas, 117; cycles of building and rebuilding, 8, 26, 44, 82; definition of, 6; development of resilient, 110–11; disaster capitalism and, 11, 16, 17; as "dynamic earth," 44; ecological footprints, 7–8, 12, 16, 21, 27, 64, 70, 84; economic development and, 10, 122; expansion of, 3, 71, 84, 115; firefighting services, 83; fire-resistant architecture, 85; global networks and, 6–7; holistic understanding of, 119; infrastructure development, 7, 47, 48; location of, 40, 117; management of financial risks, 99; mitigation of hazard, 81; multi-temporal dimensions of, 121; physical foundations of, 7, 41–42; planning, 82; population statistics, 16; profit-driven development, 11; restoration projects, 119–20; as risky places, 2; safety of, 113; sewage solutions, 45, 46; sinking of, 39, 40; spatial patterns of, 16–17, 20, 37;

value-orientation, 123; waste management, 8, 60, 61; water resources, 46–47, 50; world's 100 largest, 43
Civil Society Organizations for Disaster Reduction (GNDR), 111
Clark, Brett, 143n30
Clean Water Act, 49
Clement, Matthew, 30
climate capitalism, 108
climate change: adaptation to, 35; anthropogenic, 22–23, 55; drivers of, 7–8; food insecurity and, 81; risks of, 115; social effects of, 7; wildfire risk and, 2, 79, 81
climate data, 32
Clugston Group, 69
Cohen, Boyd, 108
Community Practitioners' Platform (CPP), 111
concrete: carbon dioxide emission and, 73–74, 85; fire-resistance of, 85
Confederation of European Waste-to-Energy Plants (CEWEP), 73
Cooling Degree Day Index, 100
CoreLogic, 32, 80
Craig, Robin, 110
Cranston Fire (2018), 9
Crassus, Marcus Lincinius, 83
creative destruction, 29, 30
Cronon, William: Uncommon Ground, 11
Cuyahoga River Fire (1969), 5

Davis, Mike, 83
Davoudi, Simin, 109
Dawson, Ashley, 40, 108
decoupling. See green growth (decoupling)
deindustrialization, 97
Department of Environment, Land, Water and Planning (Australia), 80
differential space, 15, 120–21
disaster capitalism: critique of, 91, 116, 117, 120; examples of, 36; financialization and, 12; insurance policies and, 11–12; mitigation strategies, 33, 36–37; modern cities and, 11; principles of, 4, 11, 13; profiteering and, 29; trajectories of, 19, 20
disaster funds, 101
disaster-industrial complex, 23–24, 28, 86

disaster Kuznets curve, 106
disaster management: fiscal policy and, 34–35; privatization of, 86; profit from, 31; social consequences of, 34–35; strategies for, 86
disaster mitigation, 28, 55, 81, 102, 107
disaster-oriented financial instruments, 31–32
disaster planning, 32, 106–7
disaster relief assistance, 4, 111
disaster-resilience projects, 10, 102, 110, 111, 123
disaster-resilience standards, 6
disaster-resistant architecture: development of, 25, 85, 86; wealth and, 86–87
disaster risk, 5, 8, 24
disasters: collective memory of, 119, 120; consequences of, 56; costs associated with, 2, 9; as creative destruction, 29–30; fiscal nature of, 9–10, 18, 29; gendered implications, 12; government policies and, 12; human-made, 5; as liability, 9; long-term approaches to, 105; media coverage of, 7; mitigation against, 14; physical nature of, 7–8, 18; profiting from, 4, 18, 24, 29; recovery from, 119; short-term fixes to, 3; social and political dimensions of, 2, 11–12, 17–18, 20; urban development and, 4, 20, 21
Douglas, Ian, 44
Dunlap, Riley, 41

earthquakes, 4
EarthSat, 32
East Asia: population growth, 43
ecological modernization, 75
Eco-Management and Audit Scheme (EMAS), 112
economic development: disaster-resistant technologies and, 119
Elliott, James, 30, 106
emergency loans, 101–2
energy consumption, 107
Energy from Waste (EEW), 68
Engels, Frederick, 27
environmental, social, and governance (ESG) principles, 10, 102–3, 108, 114, 116, 118, 123
environmental disasters, 17, 18

environmental Kuznets curve, 64, 106
environmental management, 75, 112
Epstein, Steven, 113
EQECAT, 32
Europe: heat waves in, 47; packaging waste in, 64
European Environmental Bureau, 97
European Insurance and Occupational Pensions Authority (EIOPA), 103, 105
European Investment Bank (EIB), 35, 102

"fail-safe" urbanism, 109
Falck company, 92
Federal Emergency Management Agency (FEMA), 76, 107
Fields, Desiree, 96, 97
finance services sector, 97–98
financial crisis of 2008, 98
financial instruments, 103
financialization, 5, 9, 13
Finland: wildfires, 78
fire. See urban fires; wildfires
Fire and Water Engineering, 23
fire codes, 87
fire ecology, 81
firefighting: budget of, 91; coordination among agencies, 90, 154n59; labor costs of, 91, 93; privatization of, 92–93, 153n45; use of inmates, 91–92, 93; use of migrant workers, 93; volunteers, 91, 151n30
Firey, Walter, 37
Flood, Joe, 83
Flood Disaster Protection Act, 51
flood insurance, 51–52
floods, 6–7, 8–9
Florida Citizens Property Insurance Corporation, 51
Florida Department of Environmental Protection, 139n44
Florida Hurricane Catastrophe Fund, 52, 53
Florida sinkholes: database for tracking, 51; hard engineering solutions for, 54–55; homeowners and, 50; insurance, 50, 51–52; legislation on, 51, 52; risk of, 51
Florida Springs and Aquifer Protection Act, 139n44
Fluor Corporation, 148n94
Fort McMurray Fire, 76, 78, 93

Foster, John Bellamy, 9, 28
Fraser, Steve, 94
Freeman, Joshua B., 94
Freudenburg, William, 120
Frickel, Scott, 106

garbage: crises, 68; examples of mishandling, 57–58; modernization of collection of, 69; as resource, 58–59
Gaz de France (GDF), 74
geoengineering, 55, 163n25
Germany: volunteer firefighters in, 151n30
GFP Emergency Services and Capstone Fire and Safety, 92
Glen Falls, NY, 45, 46
global ecology, 35
global economic loss, 19
Global Facility for Disaster Reduction and Recovery (GFDRR), 111–12
global reinsurer capital, 3, 19
Goldman Sachs, 102
Goldstein, Jesse, 108
Goodman, Philip, 91
Gotham, Kevin, 18, 30, 31
Gould, Kenneth, 108
Grayback Forestry, 92
Greece: wildfires, 78
Greenberg, Amy, 83
Greenberg, Miriam, 30
green bonds, 102, 103, 104
green cities, 108
green finance, 114
green growth (decoupling), 97, 110
greenhouse gas (GHG) emission. See carbon dioxide (CO_2) emission
Green New Deal, 108, 116, 117–18, 121, 123, 162n4
Green Star, 112
greenwashing, 95
groundwater withdrawals, 139n44
Grove, Kevin, 101
GSE Group, 69
Guatemala City: sinkholes, 43
Gunderson, Ryan, 55

Haitian earthquake, 30
hard engineering, 54
Harvey, David, 26, 119, 123
Hassler, Uta, 121

heat health warning systems, 33
Heating Degree Day Index, 100
Helvering v. Le Gierse, 98
Heraclitus of Ephesus, 77
Hoan, Daniel Webster, 48
Hong Kong: Sai Tso Wan Recreation Ground, 120
Hornbeck, Richard, 29
Hoyt, Homer, 21
Hurricane Andrew, 32, 99
Hurricane Irma, 159n43
Hurricane Katrina, 30, 31, 76
Hurricane Maria, 138n29
Hurricane Sandy, 33
Hygiene and Sanitation Company of Cameroon (HYSACAM), 74
Hyogo framework, 110

Illini Disposal, 58
incineration: as energy source, 28; problems of, 23. See also waste-to-energy (WTE)
infrastructure development, 13, 48, 58, 59, 97
Institute for Market Transformation to Sustainability (MTS), 112
insurance, 24, 29, 87, 98–99, 157n15
Inter-American Development Bank, 34
interferometric synthetic aperture radar (InSAR), 53
Intergovernmental Panel on Climate Change (IPCC), 123
International Energy Association (IEA), 72
International Monetary Fund (IMF), 9
International Red Cross and Red Crescent, 111
International Standards Organization (ISO), 5–6, 96, 112–13, 114
ISO 14000, 112, 113
ISO 31000, 6

Jakarta: Great Garuda Sea Wall project, 54; sinking of, 39
Jameson, Fredric, 16
Jasanoff, Sheila, 26
Jevons paradox, 114
Johnson, Chalmers, 35
Johnson, Leigh, 32–33, 35
Jorgenson, Andrew K., 143n30
Juvenal, 60

Kardashian, Kim, 153n45
Kellenberg, Derek, 63, 106
Kelman, Ilan, 5
Keniston, Daniel, 29
Khian Sea garbage barge incident, 62
Kim, Jeong-Chul, 24
Klein, Naomi, 4, 23, 127n7, 163n30
Klein, Richard, 110
Klinenberg, Eric, 121
Knowles, Scott, 30, 113
Knuth, Sarah, 96
Kohler, Niklaus, 121
Kunreuther, Howard, 99

Laguna Canyon Fire, 11
Lahore Waste Management Company (LWMC), 69
landfills, 23
Latvia: wildfires, 78
Leadership in Energy and Environmental Design (LEED), 96, 112, 114
Leary-Owhin, Michael, 121
Lefebvre, Henri: on humans and nature, 1, 3, 128n14; on modernity, 38; notion of differential space, 15, 19, 120; on notion of "rhythms" in social life, 18; on reality, 25; on urban space, 119, 122; on world of commodities, 118
Leitner, Helga, 112
Lewis, Tammy, 108
Liberty Mutual, 99
LifeStar Response, 92
Lin, Ken-Hou, 9
Lisbon earthquake of 1755, 4
Logan, John, 71
logistics, 68–72
London, UK: Great Fire of 1666, 5, 82, 83; Great Stink of 1858, 5, 23; subsidence-related damage, 47
Los Angeles, CA: four ecologies of, 22
Lovins, Hunter, 108
low-carbon economy, 116

Martland, Samuel, 151n30
Marx, Karl, 25, 29, 157n15
Mauch, Christof, 119
Mendocino Complex Fire, 91
Mexico City: sinking of, 39; water pipelines development, 54

Michel-Kerjan, Erwann O., 99
military-industrial complex, 24
Mill Creek, PA: sewage dumping, 45–46
Milwaukee winter, 99
mitigation, 96, 106, 107, 108, 109
Mobarak, Ahmed, 106
Mobro 4000 incident, 62
Molotch, Harvey, 71
Moore, Jason, 28
Moorpark Rubbish Disposal, 58
Morris, Corbyn, 60
mortgage insurance, 98
Mosaic Company, 39–40
Multi-Catastrophe (MultiCat) program, 101
multinational corporations, 68
Mumford, Lewis, 27, 29, 82, 115
municipal solid waste (MSW), 23, 58, 60–68, 74
Murray, Martin, 6
Musk, Elon, 32
Mutter, John, 4

National Aeronautics and Space Administration (NASA), 32
National Association of Insurance Commissioners (NAIC), 98
National Board of Fire Underwriters, 87
National Climatic Data Center, 32
National Cohesive Wildland Fire Management Strategy, 90
National Fire Protection Association, 87
National Flood Insurance Reform Act, 52
National Oceanic and Atmospheric Administration (NOAA), 32
National Pollutant Discharge Elimination System, 47, 49
National Research Council Canada, 114
National Weather Service, 32
National Wildfire Suppression Association, 92
natural disasters: assessment of threat of, 6; built environment and, 34; capitalism and, 30, 34; causes of, 105; commodification of, 31, 36; financial cost of, 2; forecasting of, 33; incident statistics, 17; logistical issues of, 8; losses from, 101; as objectively real phenomenon, 23; rhythm of, 18; risk for cities, 4–5, 20–21; sturdies of, 12

natural environment, 19, 20
nature: capitalism and, 28–29; human victories over, 27
Neely, Megan Tobias, 9
neoliberalism: firefighting strategies, 91–92; key component of, 112; urban development and, 93–94, 116; wildfire risk and, 83, 94
newly industrialized countries (NICs): ecological modernization of, 63; solid waste management in, 59, 63, 68
Newman, Peter, 163n25
New Orleans, LA: gentrification in, 31
new urbanism, 109
New York City: building codes, 87; Emergency Management department, 33; expansion over landfills, 23; fires in, 83, 87; Freshkills Park, 120; recycling in, 61–62; sinkholes, 49
Northridge earthquake (1994), 99

O'Connor, James, 28
Oliver-Smith, Anthony, 119
100 Resilient Cities initiative, 99
O'Neil, Cathy, 25
OppenheimerFunds, 103
Organization for Economic Cooperation and Development (OECD), 97
ORTEC, 69
overproduction crises, 26
Özkartallar, 68
Özpak, 69

Pacific Gas & Electric (PG&E), 1, 2, 5
Paris Agreement, 116
Pelling, Mark, 107
Pellow, D. N., 63
Petersen, Brian, 55
Pfister, Christian, 82
Philadelphia, PA: garbage site, 61
Pickett, Steward, 110
Pioneer Industries, 58
Plutarch, 83
Pollin, Robert, 162n4
Pompeii, 4
poverty: vulnerability and, 122
Price-Smith, Andrew, 35
Pyne, Stephen J., 81

Quark, Amy, 6
Quinn, Sarah, 9

Rafferty, Michael, 103
Rangeland Fire Protection Associations, 90
Raytheon Company, 32
Recuber, Timothy, 41
recycling, 28
Re:Focus firm, 102
regenerative urbanism, 118, 119, 120, 121, 124
RELi, 6, 112, 114
renewable energy, 10
Republic Services company, 58
resilience: bonds, 102, 105; definition of, 96, 110, 111; social actors and, 111–12; social capital and, 121; standards, 112–13
resilient cities, 14, 15, 110–12, 113
rhythms of daily life, 18, 19
Rice, James, 71
right to generate, 163n30
risk, positive and negative, 6
risk assessment mechanisms, 96
RiskLink, 32
risk management: Big Data and, 25; capitalism and, 95–96; language and practice of, 6–7; neoliberal policymaking and, 30; profiting from, 5–6, 31, 98; by public agencies, 33; reliance on market, 95–96; strategies for, 107; urban governance and, 23
Risk Management Solution (RMS), 32
risk reduction, 96–97, 98
risk society, 5
Rogers Helicopters, 92
Roman cities: drainage systems, 45; sewers in, 60
rubbish, 58

San Francisco, CA: earthquake of 1906, 4, 83; Millennium Tower, 44
Schnaiberg, Allan, 10, 63, 70, 75
Schumpeter, Joseph, 29
Scott, Andrew, 77
Seattle, WA: Ravenna sinkhole, 49
Sendai framework, 110
Sennett, Richard, 21
sewage: disposal of, 45–46, 58; growth of, 42; regulation of, 46–47; sinkholes and, 47

shack fires, 83
Shanghai Chengtou Holding Co., Ltd., 68
Shanghai Environment Group (SEG), 58, 68
shelter-in-place (SIP): implementation of, 88
Sherlock, Robert Lionel, 44
shock doctrine, 127n7
Siberian wildfires, 78
Sicotte, Diane, 61
sinkholes: causes of, 39, 49, 56; in China, 42–43; classification of, 46; cost of fixing, 49; damage caused by, 49; as disposal sites, 45; formation of, 38–39; human activity and, 39–40; in Japan, 43; meanings of, 44–45; as metaphor, 48; mitigation strategies, 49; natural, 44; pipe systems and, 138n29; prediction of, 53, 56; public perception of, 7, 41; risk of, 13, 47, 53; sewer systems and, 47, 55; in South Africa, 43; urban, 13, 38–39, 42; in the US, 42; for water disposal, 46
sinking cities, 39, 40–42
Skidmore, Mark, 30
smart cities, 10, 25–26, 97, 109, 117
"smart growth," 109
solid waste management, 45, 62, 63, 64–65
Sorensen, J. H., 90
South Africa: fynbos vegetation, 84; informal settlements, 80; shack fires in, 80; waste management in, 68; wildfires in, 77
space: social production of, 19
SpaceX, 32
Spirn, Anne, 22
Squires, Gregory, 98
Starco Demarco, 65, 67, 75
start-ups, 33
State College, PA, 45
Steinbeck, John, 57
Steinberg, Ted, 79
Strange, Susan, 33
Stuart, Diana, 55
Suez GDF/Engie, 74
sustainable development, 96, 97, 109, 110
Swart, Rob J., 14
Sweden: wildfires, 78
Swiss Reinsurance Company (SwissRe), 47, 99, 101, 102

172 • Index

Tampa Bay Metropolitan Statistical Area, 50
technological disasters, 17
technology, 25, 124
10 Tanker LLC, 92
Thompson, Michael, 58
Thomson, Giles, 163n25
Tianying Inc., 68
Tierney, Kathleen, 119
Timmermans, Stefan, 113
Toya, Hideki, 30
trash: commodification of, 59–60; definition of, 58; incineration of, 23, 78, 133n47; negative impact of, 59; pollution due to, 23
Travelers, 99
treadmill of production, 70, 71, 75
Tropical Storm Agatha, 43
Turbeville, Daniel, 82
Turkey: urban growth, 69–70; wildfires, 78

Underground Injection Control (UIC) Class V Wells ("improved sinkholes"), 46
United Nations: relief aid, 111; resilient cities initiative, 110–11; risk management policies, 34
United Nations Climate Change Conference in Paris (COP15), 10
United Nations Development Programme (UNDP), 110
United Nations Educational, Scientific, and Cultural Organization (UNESCO), 41
United Nations International Strategy for Disaster Reduction (UNISDR), 109, 110, 111, 113
United Nations Joint Environmental Unit, 75
urban development: anthropocentrism of, 21; building codes and, 87; capitalism and, 18, 22, 116; catastrophes and, 30, 31; climate and, 22–23; fires and, 77, 78–79, 80, 81–82, 93–94; funding of, 118; geography and, 22; in high-risk areas, 80; neoliberalism and, 21, 93, 116; reliance on Big Data, 25; short-term solutions, 35; zoning practices, 93–94. See also regenerative urbanism

urban disaster, 23, 35; risk of, 11, 20–21, 114
urban environment, 96–97
urban fires: causes of, 82, 85; disaster capitalism and, 86; history of, 81–84; mitigation strategies, 82, 85–86; profiting from, 82–83; urban development and, 77, 78, 83, 87; vs. wildfires, 83
urban forests, 124
urban governance, 123–24
urban growth, 71, 79, 80, 81, 84
urbanization, 3–4, 5, 27
urban metabolism, 42
urban nature, 21, 22
urban planning, 64–65, 122
urban precarity, 33
Urbaser-Danner company, 65, 75
U.S. Army Corps of Engineers, 75, 76
U.S. Department of Agriculture, 90
U.S. Department of the Interior, 90
U.S. Environmental Protection Agency (EPA), 8, 75, 128n17
U.S. Forestry Service, 89, 90, 92
U.S. Green Building Council (USGBC), 6, 107–8, 112

Vale, Lawrence, 111–12
Van der Knaap, Wim G. M., 14
Venice: sinking of, 39
Veolia, 62
vulnerability, 122

Wacquant, Loïc, 91
waste: collection of, 61, 75–76; commodification of, 70, 71, 72; as construction material, 23; consumption ideology and, 14; disasters and, 8; as hazard, 72; movement of, 62–63; risks associated with, 13–14, 23; studies of, 64
waste collection workforce, 61, 141n14
waste crises: global nature of, 64
Waste Group, 68
waste management: early efforts of, 23; firms, 58; global, 64, 65, 66, 68; green approach to, 105–6; logistics of, 59, 68–72; municipal responsibility for, 62; outsourcing of, 62; policies, 8, 13–14; privatization of, 65; profitability of, 58. See also solid waste management
Waste Management, Inc. (WM), 58, 75

waste-to-energy (WTE): in Asian countries, 72; companies, 74; development of, 60, 61; disaster capitalism and, 72–75; in Europe, 73; financialization of, 28; green, 73; logistics of, 73, 74; marketing of, 73; profitability of, 72, 73; regulation of, 61
water: consumption of, 107; contamination of, 8–9; problem of absence of, 47
weather derivatives, 99, 100
weather exchanges, 32
weather forecasting, 33
weather patterns management, 99
Weinberg, A., 63
West, Kanye, 153n45
Whitnall, Gordon G., 87
Wildfire Defense System, Inc., 86
wildfires: capitalism and, 81; causes of, 78; climate change and, 79, 81; cost of, 78; as global issue, 77–78; harm from, 79, 80, 90; human activity and, 1, 14, 81–82, 84; natural vegetation and, 84; political ecology of, 14; social consequence of, 78, 79; studies of, 84; urban development and, 2, 11, 14, 78–79, 80, 80
wildland–urban interface (WUI): benefits and risks of, 22, 36
Wirth, Louis, 21, 23
Woolsey Fire (2018), 9
World Bank, 34, 102, 111
World Health Organization (WHO), 75
World Trade Organization (WTO), 63
Wren, Christopher, 82

Yokohama strategy, 110

Zimring, Carl, 45

About the Author

ALBERT S. FU is professor of sociology at Kutztown University of Pennsylvania. He has previously published articles in *Cities, City & Community, Critical Sociology, International Journal of Urban and Regional Research,* and *Urban Studies.*

Available titles in the Nature, Society, and Culture series:

Diane C. Bates, *Superstorm Sandy: The Inevitable Destruction and Reconstruction of the Jersey Shore*

Soraya Boudia, Angela N. H. Creager, Scott Frickel, Emmanuel Henry, Nathalie Jas, Carsten Reinhardt, and Jody A. Roberts, *Residues: Thinking through Chemical Environments*

Elizabeth Cherry, *For the Birds: Protecting Wildlife through the Naturalist Gaze*

Cody Ferguson, *This Is Our Land: Grassroots Environmentalism in the Late Twentieth Century*

Albert S. Fu, *Risky Cities: The Physical and Fiscal Nature of Disaster Capitalism*

Shaun A. Golding, *Electric Mountains: Climate, Power, and Justice in an Energy Transition*

Aya H. Kimura and Abby Kinchy, *Science by the People: Participation, Power, and the Politics of Environmental Knowledge*

Anthony B. Ladd, ed., *Fractured Communities: Risk, Impacts, and Protest against Hydraulic Fracking in U.S. Shale Regions*

Stefano B. Longo, Rebecca Clausen, and Brett Clark, *The Tragedy of the Commodity: Oceans, Fisheries, and Aquaculture*

Stephanie A. Malin, *The Price of Nuclear Power: Uranium Communities and Environmental Justice*

Stephanie A. Malin and Meghan Elizabeth Kallman, *Building Something Better: Environmental Crises and the Promise of Community Change*

Kari Marie Norgaard, *Salmon and Acorns Feed Our People: Colonialism, Nature, and Social Action*

J. P. Sapinski, Holly Jean Buck, and Andreas Malm, eds., *Has It Come to This? The Promises and Perils of Geoengineering on the Brink*

Chelsea Schelly, *Dwelling in Resistance: Living with Alternative Technologies in America*

Sara Shostak, *Back to the Roots: Memory, Inequality, and Urban Agriculture*

Diane Sicotte, *From Workshop to Waste Magnet: Environmental Inequality in the Philadelphia Region*

Sainath Suryanarayanan and Daniel Lee Kleinman, *Vanishing Bees: Science, Politics, and Honeybee Health*

Patricia Widener, *Toxic and Intoxicating Oil: Discovery, Resistance, and Justice in Aotearoa New Zealand*